Clerk of Works and Site Inspector Handbook

2018 Edition

Clerk of Works and Site Inspector Handbook

2018 Edition

ICWCI
Founded 1882

The Institute of Clerks of Works
and Construction Inspectorate
of Great Britain Incorporated

RIBA Publishing

© The Institute of Clerks of Works and Construction Inspectorate of GB Inc. and RIBA Publishing 2018
Previous edition 2006
Published by RIBA Publishing, 66 Portland Place, London, W1B 1NT

ISBN 978-1-85946-711-4

The right of the Institute of Clerks of Works and Construction Inspectorate of GB Inc. to be identified as the Author of this Work has been asserted in accordance with the Copyright, Designs and Patents Act 1988 sections 77 and 78.

British Library Cataloguing-in-Publication Data
A catalogue record for this book is available from the British Library.

Commissioning Editor: Alex White
Project Editor: Daniel Culver
Production: Jane Rogers
Design: based on a design by Ben Millbank
Typeset by Fakenham Prepress Solutions, Fakenham, Norfolk NR21 8NL
Printed and bound by Page Bros, Norwich
Cover Image: © By yuttana Contributor Studio / Shutterstock.com

While every effort has been made to check the accuracy and quality of the information given in this publication, neither the Author nor the Publisher accept any responsibility for the subsequent use of this information, for any errors or omissions that it may contain, or for any misunderstandings arising from it.

www.ribapublishing.com

Contents

Contents

Foreword

Eyes, ears and action.

With the ever-increasing backdrop of a strong construction pipeline, particularly for homes and infrastructure, and the need to build faster and with modern methods of construction, there has never been a more important time for the Clerk of Works role on projects.

Catastrophic events, such as Grenfell Tower, will forever be poignant reminders of the need for those critical eyes and ears on site as construction progresses, as well as the fundamental importance of professional competence.

There's an increasing need for public trust in our professions and recognised competence provided through professional qualification, as well as high standards and regulations. All of these play to the strengths and command for driving action through the Clerk of Works. As many other professions reach key milestones in 2018, such as the Institution of Civil Engineers in its 200th year and the Royal Institution of Chartered Surveyors in its 150th year, it is a timely moment to recognise and acclaim the benefits brought to historical construction delivery by the The Institute of Clerks of Works and Construction Inspectorate (ICWCI) in over 130 years of history. But now is not the time to look back, more to look forward, with 'ability, integrity and vigilance'. It is as vital now and for the future, if not more so, than it was in the 19th century.

For quality control on, and off, site my mantra is 'long live the Clerk of Works'.

Amanda Clack, Executive Director at CBRE and Past President at RICS

Acknowledgements

The Institute of Clerks of Works and Construction Inspectorate wishes to place on record its thanks and appreciation to the following individuals who have played such a vital role in the development of this latest edition of the handbook. The team involved with this revised version of the handbook were: Rachel Morris Hon. FICWCI, Ian Carey FICWCI, Jerry Shoolbred FICWCI, Charles Lowrie FICWCI, David Gray FICWCI, John Ruddell FICWCI, Ron Philpot MICWCI, Dr Paul Hampton FICWCI, Phil Singleton FICWCI, Derek Bishop Hon. FICWCI, Marcus Millet MICWCI, Neil Dickinson FICWCI, Spencer Henry MICWCI (Electrical) Mike Oxford CEcol, FCIEEM, James Burgoyne LLB ACII, Deborah Stead (BSi), LABC Warranty.

Introduction

Ability, integrity and vigilance

One of the most common misconceptions, certainly among the uninitiated, is the belief that a Clerk of Works' background and experience is of a purely secretarial or administrative nature. The title has been around since the Middle Ages, and it has been worn with great pride by many thousands of exponents of the craft since that time.

Although it is difficult to confirm an exact date, the title is traceable as far back as the 13th century when the Church (the richest single entity in the British Isles at the time) was undertaking a vast amount of construction work. It is believed that materials were being pilfered or otherwise lost to such a degree that some form of control over quantities and their ultimate usage was required — and the role of Clerk of Works was born. Initially drawn from the ranks of the clerics in holy orders (believed to be the most literate and honest of all) with responsibility for the Church works, they were deemed to possess the essential attributes of 'ability, integrity and vigilance'. The original title 'Cleric of the Church Works' soon became shortened to 'Clerk of Works' and remains to this day. One of the earliest-known Clerks of Works, although not the first, was the poet Geoffrey Chaucer, who was appointed as 'Clerk of the King's Works' from 1389 to 1391, superintending undertakings in Woolwich and Smithfield in London.

Over the centuries, the role and duties of the Clerk of Works developed as a part of the great historical and social changes, and exponents were selected from suitable building craftsmen, with proven technical experience, knowledge and personality gained over many years 'on the tools'. The Reform Bill of 1832 and the many Acts that followed it covering education, public health, roads, railways, town improvements and industrial development spawned many technical, professional and scientific associations and institutes.

On 10 March 1882, a group of 17 Clerks of Works met and agreed that they would form a professional body named 'The Clerks of Works Association'. At their first general meeting, held on 27 March 1882, and with 31 present, they formed their first governing body, electing Mr Thomas Potter as their first President, issuing their first rulebook and extending the formal title of the association to 'The Clerks of Works Association of Great Britain'. In 1903 the Association gained incorporated status and was again retitled, 'The Incorporated Clerks of Works Association', which remained in place until 1947, when it became 'The Institute of Clerks of Works of Great Britain Incorporated' (ICW). Finally, in 2009, the name was changed to 'The Institute of Clerks of Works and Construction Inspectorate of Great Britain Incorporated' (ICWCI) to encompass and encourage into membership professionals who practise in construction inspection specialisms, while still including those who continue in the traditional role of Clerk of Works.

It is important to realise that although building practices and procedures have come and gone, and come again in some cases, over the many years since ICWCI in its original state was formed, the very principles for which Clerks of Works were first established remain as true and valid as ever. Tragically, in many cases the traditional title has given way to a more modern interpretation of the role – construction inspector, site inspector, quality auditor – but the principle remains the same. Whether it be in the field of civil engineering, building, landscaping, tunnelling, electrical and mechanical engineering, new build or refurbishment, and in the UK or elsewhere, the 'Inspectors' ply their craft in ensuring value for money for the client, through rigorous inspection of the materials in use, and the craftsmanship deployed in their usage. Their skills are honed not only from books and study, but also from years of practical, hands-on experience. They know, or will have seen, almost every trick in the book, and will know equally well how to counter it – there being no better gamekeepers than, of course, former poachers! They must remain up to date in terms of legislation and the many building regulations, including those that relate as much to the individual – health and safety, race relations, disability discrimination – as to the act of construction itself. Through rigorous and detailed reporting and record keeping, and thorough inspection of specifications and drawings, their work will, without question, add value to any project – even though it may not be obvious at the time. The question that should always be asked is a very simple one:

'Without the intervention of the Clerk of Works – how much would rectification and/ or remedial action have cost?'

After all, the motto of the Clerk of Works is 'ability, integrity and vigilance'!

The late Peter Lennon FICWCI
Past President and Chairman of the ICWCI Professional Standards Committee

Appointing a Clerk of Works

1.1 APPOINTING A CLERK OF WORKS

A Clerk of Works can provide professional services in five main ways.

1. **As a full-time or part-time 'permanent' employee in what the law terms 'a master-servant relationship'.** Such a person might be resident on one project or required to carry out duties as a visiting Clerk of Works on several projects. There will be a contract of employment, and taxation under PAYE. The employer will normally be vicariously liable for the actions of the Clerk of Works in the course of employment.
2. **As an employee on a fixed-term basis taken on to perform specific tasks under a fixed-term contract of employment.** Such a person will be under the control of the employer in a master-servant relationship. There will be a contract of employment and taxation under Schedule E (PAYE).
3. **As a director of a company,** still technically an employee of the company. The contract for the Clerk of Works services will be between the client and the company.
4. **As a sole practitioner, i.e. an independent person in business on their own account offering professional services.** Such a person will perform specific services in return for a fee. There will be a contract of engagement, and taxation under PAYE. Public liability (PL) insurance and professional indemnity (PI) insurance will normally be required.
5. **As a partner in a firm.** Partners will be jointly liable with their other partners for their acts and omissions. The partners in the firm will enter into a contract of engagement to perform specific services in return for a fee and be subject to appropriate taxation. PL and PI insurance will normally be required.

1.2 SELECTION AND INTERVIEW

1.2.1 Factors affecting selection

Various factors are likely to influence the selection of Clerks of Works with the appropriate skills and experience, and definition of a proper level of remuneration and conditions of appointment. Consideration should be given to the following:

- the nature of the building project (e.g. new build, refurbishment, engineering, landscaping)
- the nature of the services required (e.g. traditional construction with recognised basic trades and skills, predominantly specialist with a high content of advanced structural work, or with a high content of specialist subcontractor work)
- the approximate scale of the project and the expected length of the appointment
- the status and level of responsibility envisaged (e.g. as sole construction inspector, or as part of a team of Clerks of Works) and the degree of accountability
- the type of procurement path for the project
- the type of document used for appointing the Clerk of Works.

1.2.2 Pre-interview submission

For the appointment of a sole practitioner or a firm of Clerks of Works, a submission before interview might be called for, and therefore there might be a requirement for a free quotation in competition with others. Sometimes interviews are held first and the preferred applicant is invited to negotiate terms with the appointing body subsequently. A submission might include the following information:

- name, location and form of practice, and how long it has been established
- names and qualifications of key staff, and range of skills available
- details of a minimum of four recently completed appointments (e.g. project name, description, procurement method, cost, programme, with name of client and lead consultant)
- details of current appointments and present workload
- details of experience in matters relevant to the proposed appointment (e.g. quality management/assurance, quality control, site testing methods)
- details of financial standing (including name of bankers) and evidence of satisfactory professional indemnity insurance, where appropriate.

1.2.3 ICWCI membership

For all appointments, employers are recommended to take account of membership of the ICWCI as offering an endorsement of competence and an indication of an appointee's interest in continuing professional development (CPD). For larger contracts, or those that entail a higher level of responsibility, it might be reasonable to require corporate membership. This would ensure a degree of experience gained at an appropriate level.

Interviews for salaried staff to be employed by an authority or organisation will follow the pattern and procedures used by the particular organisation.

1.3 APPOINTMENT

1.3.1 Appointment documentation

Appropriate terms and conditions of appointment for freelance Clerks of Works are published by the ICWCI. There is a memorandum of agreement between the client and the Clerk of Works that includes directions for fees on a lump sum, percentage or time charged basis, and a separate supporting schedule of services and fees, available from ICWCI subject to invoice. The memorandum incorporates by reference the conditions appropriate to the appointment, and these are published separately in the *Clerks of Works/Construction Inspector Appointment Document* (which can be purchased by contacting ICWCI Headquarters). This includes sections covering the following:

1. Services pre-construction (prior to the start of the building contract)
2. Services pre-construction (provided at the start of the building contract)
3. General services – construction stage
4. Services post-construction
5. Other services
6. Conditions of appointment
7. Fees and expenses.

There are separate helpful guidance notes contained in this document; some clients may welcome professional advice on which of the sections should apply in particular circumstances.

1.3.2 Letter of appointment

The appointment of a Clerk of Works should always be confirmed in a properly formulated letter of appointment that sets out the terms of the agreement between the employer and employee and should include:

- a description of the project, including site layout, etc.
- the conditions on which the appointment is based
- the scope and duration of the appointment
- details of remuneration, and the method of calculating fees, expenses, etc.
- the identity of client representatives and of other appointed consultants
- arrangements for professional indemnity insurance.

An employer must give employees a 'written statement of employment particulars' if their employment contract lasts at least a month. This isn't an employment contract but will include the main conditions of employment. The employer must provide the written statement within two months of the start of employment.

The latest guidance and updates in relation to employment, contracts and working in another county can be found on the government website (www.gov.uk).

Role and Relationships

2.1 THE ROLE OF THE CLERK OF WORKS

Notwithstanding the role of the Clerk of Works being specifically mentioned within various bespoke forms of contract, a pragmatic definition, which is still highly relevant today, would be, as paraphrased, in the old Greater London Council *Handbook for Clerks of Works*:

'The role of the Clerk of Works can be defined as being the representative of either the Client, CA [contract administrator] (or relevant stakeholder) on site, responsible for the detailed inspection of works in progress to ensure that such works are executed in accordance with the contract documents and any instructions that may be issued from time to time.'

'The Clerk of Works should ensure that the specified standard of workmanship is maintained, that the materials and products used are of the specified quality, that construction throughout is sound, that the progress of the work accords with the contract requirements, and that all essential facts relating to the work are properly recorded.'

2.2 THE FUNCTION OF THE CLERK OF WORKS

The National Occupational Standards (NOS) for Site Inspection (Construction L3) identifies five core or mandatary competencies that a capable and competent site inspector/Clerk of Works must adequately demonstrate and perform, i.e. monitor projects, monitor health and safety systems and responsibilities, prepare documentation for handover and obtain project feedback, prepare and organise site inspection information, and practise in a professional and ethical manner.

So, in general terms, the functions and attributes required of a professional Clerk of Works are as follows.

- **Anticipation:** the ability to identify problems in advance to prevent them materialising or, where they do materialise, to help overcome them quickly. Intelligent anticipation is founded on experience and a thorough knowledge of the contract documents. A competent Clerk of Works functions as an early warning system.
- **Interpretation:** verifying that the contractor fully understands instructions given in words and on drawings and acting to remove ambiguity.
- **Recording:** making as complete a record as is appropriate, bearing in mind the heavy reliance that may be placed on the accuracy and objectivity of that record by the architect and the others.
- **Inspection:** detecting workmanship that does not comply or materials that do not conform to the contract standards. This will usually mean inspecting in detail, and verifying measurements regularly.

- **Reporting:** keeping the architect fully informed on a regular pre-arranged basis. This also means alerting the architect immediately when situations arise that require decisions or actions.

In a traditional building contract, the functions described above will involve the Clerk of Works in the monitoring activities described in sections 2.3 to 2.5 below. It should be noted that the term 'architect' should be taken to include any other consultant named as administering the contract.

2.3 MONITORING QUALITY CONTROL ON AND OFF SITE

It is a contractual obligation for the contractor to produce work and use materials in accordance with the standards of the building contract. Verification that this obligation is being met is seldom subject to formal procedures. It should be made clear at the outset whether the Clerk of Works is to act as a quality checker or controller, and agreement should be reached about the methods to be adopted. Where there is to be a programme of predictive inspection, the Clerk of Works will need an inspection plan, a quality plan and a test plan.

Where the contractor is required by the employer to apply quality management methods under BS EN ISO 9001:2015, the Clerk of Works should ensure that the agreed method statement and quality plan for the project are compatible and are being observed, that the quality audits are carried out as programmed and that non-conforming elements are dealt with expediently.

2.4 MONITORING THE ADEQUACY AND FLOW OF PROJECT INFORMATION

Incomplete and allegedly late drawings are a serious problem on many projects. The Clerk of Works can make an important contribution by anticipating when certain information will be needed and notifying the architect without delay. Contractors might be required to produce a schedule of information showing what they still need and when it will be needed. They will normally present this at the initial project or pre-start meeting, where it might be accepted or amended by the architect.

The Clerk of Works should review drawings and other information at or prior to issue and draw the architect's attention to inconsistencies in and between documents. Clerks of Works are sometimes required to interpret drawings for the contractor, and may be able to supply minor items of missing information, although they should be careful not to exceed their authority or incur unauthorised expenditure or liability.

The Clerk of Works has an important role in assisting with the compilation of record and maintenance information, and, where the contract requires it, in contributing information for 'as-built' drawings. On larger and more sophisticated projects, this might call for a Clerk of Works to access information held on a dedicated portal, project extranet or other information system.

2.5 MONITORING THE EFFECTIVENESS OF SITE MANAGEMENT

Efficient site organisation by a contractor is evidenced by a firm management policy, an effective operational structure and clear lines of authority and accountability. Some contractors, particularly those selected by negotiation, are appointed because of their good track record in site management, but on traditional building contracts there is often considerable variation in management performance. In this respect, the Clerk of Works can have an important monitoring function.

Where a contractor's site operations are subject to quality management procedures, the Clerk of Works has a clearly defined framework for monitoring them. However, where they are more ambiguous, a vigilant Clerks of Works will guard against becoming implicated beyond their authority. First-hand knowledge and excellent records are essential; it might become necessary later to show that regular progress was possible but was inhibited by the contractor's inadequate management.

Monitoring the contractor's obligation to comply with statutory requirements is likely to bring Clerks of Works into direct contact with building control officers, approved inspectors and other representatives of statutory bodies who visit the site. They will inform the architect immediately of matters raised and agreements reached, and will subsequently monitor the contractor's compliance. Health and safety regulations are having an increasing impact on the construction industry, therefore the Clerk of Works should be fully conversant with the Health and Safety Plan required under the Construction (Design and Management) Regulations (CDM 2015) and should monitor the contractor's compliance and draw any deviations to their attention.

2.6 WORKING RELATIONSHIPS

2.6.1 With the employer under the building contract

Many Joint Contracts Tribunal (JCT) building contracts refer to the Clerk of Works as acting solely as an inspector on behalf of the employer under the direction of the architect/contract administrator, who is acting as an agent of the employer. In the traditional situation, where the Clerk of Works has been appointed on a salary basis and is effectively an employee of the employer, the latter is generally vicariously liable for the Clerk of Works' actions.

The employer's interests can be protected in many ways before the works start by doing a conditions survey or dilapidations survey. For example, the Clerk of Works should immediately make a careful record, including photographs, of the condition of pavements, fences and points of site access. Any allegations by adjoining owners or instances witnessed of damage due to the contractor's operations should be reported immediately.

The primary responsibilities of Clerks of Works will arise from their contracts with those who employ them. Clerks of Works are almost invariably employed and paid by the building owner and are intended to assist the architect in the discharge of the architect's duties of supervision and control of the work. However, while this will usually reflect their sole contractual responsibilities and liabilities, Clerks of Works may possibly be liable to third parties for loss resulting from negligence under the laws of tort.

2.6.2 With a project manager or employer's agent

Many projects are now administered by a project manager (PM), particularly under an NEC form of contract. The RIBA has identified the PM as being a client's representative providing independent management of a project, from identification of business need to completion. In some procurement arrangements the PM is the supra-consultant placed above the design team, consultants and contractor, to whom they are all directly accountable. The term is not recognised in JCT contracts, which refer to the architect/contract administrator, but there is nothing to prevent the PM from carrying out this role. However, government construction contracts (e.g. GC/Wks/1 edition 3) refer to the PM as a named person appointed for the purpose of managing and superintending the works. In this context, the PM is permitted to delegate any powers and duties, and a Clerk of Works might be appointed to exercise many of these duties.

The term employer's agent (EA) is used to describe an individual (or organisation) acting on behalf of the client as the contract administrator under JCT design and build forms of contract. They may be a lead consultant, often the quantity surveyor or the architect, or in some circumstances, they may be a member of the client's in-house team.

Where a PM or EA has been appointed, it is essential to establish in what capacity they will operate. The relationship between the PM/EA and the Clerk of Works and the procedures that are to be adopted should be clarified. The Clerk of Works might still work under the direction of the architect or any other consultant named as administering the contract, but there might be some differences in working methods: for example, the intervals at which consultants' reports are to be received (usually monthly), any preferred structure for meetings, specific information to be included in progress reports, particular checklists to be used at practical completion, commissioning and handover, etc.

2.6.3 With the architect

The JCT Standard Building Contract (SBC) refers to the Clerk of Works acting on behalf of the employer under the direction of the architect. This is the conventional situation where the Clerk of Works has a watching and recording brief and is the eyes and ears of the architect. A period working with the architect before going to site can prove invaluable in establishing the basis for a successful working relationship.

The effectiveness of the Clerk of Works will depend to a large extent on a good, clear and complete briefing at the outset, on the adequate and timely flow of project information, and on the architect's experience as a contract administrator.

As an inspector on site, the Clerk of Works will inform the architect immediately it is believed that the work deviates from the contract requirements. Although working under the control of the architect, Clerks of Works are ultimately accountable to their employer, the building owner. Some building contracts empower the Clerk of Works to issue directions (using the Clerk of Works Direction Form – see pp195–96) to the contractor subject to their confirmation as instructions by the architect/contract administrator; some others might confer greater power, including the issuing of instructions to the contractor. In these circumstances, Clerks of Works would be acting as agents, and would thereby attract more onerous liabilities than those attached to their traditional role. Much will depend on the authority given at the outset, and on the wording of the particular contract and cover under the architect's PI insurance.

In any event, the Clerk of Works should record and report progress, conditions, events, deviations and any matters that may lead to possible future problems. The architect will rely heavily on the Clerk of Works' anticipation of events, frank reports and helpful suggestions. The relationship between architect and Clerk of Works is founded on mutual trust and respect; its effectiveness depends on the Clerk of Works having no doubt about the architect's expectations, being given the appropriate delegated authority and possessing the will to use it.

On some contracts a resident site architect is appointed. In this event, the role of the Clerk of Works in relation to the site architect will need to be clearly defined. It should be established whether reports are to be made to the latter or to a named person at the architect's permanent offices.

2.6.4 With consultants

Quantity surveyor

The quantity surveyor named in the JCT SBC is responsible for measurement, valuation and adjustment of valuations. The Clerk of Works can assist by:

- making a detailed record of items of work before they are covered up
- making a record of all deliveries to site
- keeping a record of work carried out on a day work basis, and verifying that such work has been carried out.

Day work sheets should be signed 'for record purposes only', since it is the architect together with the quantity surveyor who will decide whether or not the work is to be paid as day work.

Engineering consultants

Although structural and building services consultants do not usually have any executive power under the JCT SBC, they have an important inspection role. Any instructions that consultants might wish to pass to the main contractor or relevant subcontractor must be channelled through the architect, and it is important for the Project Clerk of Works to be kept fully informed of any instructions issued.

The structural consultant will be concerned that the design performance criteria are met in the construction methods, materials and workmanship, both in off-site fabrication and on-site assembly. The Clerk of Works may be involved in verifying tolerances, joints, reinforcement positions, concrete pouring operations and finishes. In each instance, clear instructions from the consultant will be needed. Regular verification of materials (e.g. aggregate) and tests (e.g. piles, slump, concrete cube, reinforcement, mortar) may need to be planned and observed, and the results recorded. The Clerk of Works will maintain correct working relationships with both the consultant and the architect, bearing in mind that the latter has the executive responsibility.

The services consultants will normally be responsible for the inspection, integration and coordination of services systems, and for their subsequent installation, testing, commissioning and planned maintenance. Services installations are often complex and the documentation may be difficult for the non-specialist to understand. The Clerk of Works may have to rely heavily on effective communication with the relevant consultant. As work proceeds, it is critical for builders' work and the routing of pipes and cables to be correctly executed. A programme of progressive inspection on a component, sub-assembly and section basis might be needed and verification implemented. The Project Clerk of Works should make sure that, before services installations work is covered up, the architect is advised and given the opportunity to inspect the completed work.

The Clerk of Works may be required to witness testing, balancing and commissioning, as provided for in the subcontract documents. These operations need to be effectively integrated into an overall programme to avoid delays to completion. Where these considerations are within the remit of specialist consultants (and sometimes Services Mechanical and Electrical [M&E] Clerks of Works), integration becomes even more exacting. The Services Clerk of Works should also maintain a cooperative working relationship with any specialist on site, and inform the services engineer whenever an instruction from a consultant to a subcontractor seems necessary.

In assisting consultants, the services Clerk of Works may:

- coordinate requirements and information
- if requested, make practical suggestions and provide information
- arrange for tests required under the contract to be carried out and observed, and record and report the results.

2.6.5 With the contractor

The Clerk of Works should, as necessary, remind the contractor or 'person in charge' (PIC) of the standard of quality required under the contract, dealing promptly and firmly with any departure from good practice or disregard of architect's instructions. However, good cooperation between the Clerk of Works and the contractor's agent can contribute significantly to the success of the contract, particularly in ensuring that the site is kept in an orderly fashion, unfixed materials are properly stored and completed work is adequately protected. The Clerk of Works should also work closely with the contractor to see that tests, inspections and preparations for formal handover are all effectively supervised.

The Clerk of Works will be particularly vigilant on site in the following respects:

- in inspections, as programmed or as otherwise recognised
- in carrying out tests, as programmed or as otherwise required
- in agreeing measurements prior to work being covered up
- in providing directions or instructions and monitoring work in progress.

2.6.6 With other Clerks of Works

When more than one Clerk of Works has been appointed it is important that one is nominated and recognised as the Project Clerk of Works. This is the person who will receive instructions from the architect, submit reports and attend the site progress meetings for the Clerk of Works items. Only the Project Clerk of Works should be empowered to issue directions under the JCT SBC. A good working relationship with other Clerks of Works is essential, and with any other site supervisory and management staff. It will be important to maintain a consistent interpretation of the situation on site and a coordinated approach towards monitoring, inspection, testing, recording and reporting.

Duties of the Clerk of Works/ Site Inspector

3

This section provides a general description of the duties of a Clerk of Works. However, it should be noted that these duties will vary depending on the type of procurement and form of contract for each particular project.

3.1 THE CLERK OF WORKS' DUTIES

There are many advantages in appointing a Clerk of Works before work starts on site. This arrangement may prove more difficult in the private sector, but the Clerk of Works can often comment from experience on construction matters and make a significant contribution to production information. A period of two weeks spent in the architect's office before going on site will enable a Clerk of Works to get to know the personnel, procedures and documentation associated with the project.

3.1.1 Pre-project preparations

During this period of familiarisation, the Clerk of Works might be expected to:

- study the architect's office procedures
- study the contract drawings and schedules, and information produced by consultants and specialists
- study the programme for issue of further information
- study bills of quantities and/or specifications
- study the Health and Safety Plan for the project
- study relevant British Standards and Approved Codes of Practice, etc.
- establish lines of communication with the project architect, quantity surveyor, other consultants, other Clerks of Works and the contractor's site supervisory staff
- clarify with the project architect the standards to be met under the contract
- as requested, advise on materials, construction details, samples, etc.
- assemble the record documents supplied for use on site and clarify how they are expected to be used
- as relevant, make contact with the local authority or other approved inspectors
- make arrangements for obtaining weather information
- visit the workshops for proposed nominated suppliers and subcontractors as appropriate and necessary.

3.1.2 Documentation

The Clerk of Works might expect to be supplied with the following documents:

- form of building contract, incorporating any supplements or amendments
- contract bills of quantities (unpriced) and/or specification
- 'numbered documents' or other information relating to subcontracts
- employer's health and safety policy requirements and the approved Health and Safety Plan

- site diary
- report forms
- quality management plan, contractor's method statement, quality control checklist, verification forms, etc.
- relevant record forms
- record chart for rainfall, etc.
- copies of instructions applicable to the project.

Site accommodation provided for the Clerk of Works should be verified to see that its position and siting, furniture and equipment are in accordance with what has been specified.

3.1.3 Briefing

Clerks of Works should be briefed about their responsibilities and extent of authority in connection with:

- hours of working and notification of additional hours
- daily labour returns and method of submitting them
- method of recording time lost in site working
- signing of authorised day work vouchers
- samples of materials
- testing of samples
- storage of materials
- notification of work to be covered up after inspection
- general procedures for inspection and recording
- visitors to site and permission to take photographs
- the Party Wall Act 1966 (note if applicable).

Site supervisory staff might be briefed at a special meeting called by the architect or as part of the initial project or pre-start meeting. Ideally, the full project team will be present so that all personnel can be introduced and identified, and responsibilities and lines of communication defined and established.

3.1.4 Initial project or pre-start meeting

The agenda for the meeting will include an item designated 'Clerks of Works Matters', covering inspection, facilities and quantity verification. The architect, who will chair the meeting, will clarify the anticipated pattern of visits to site and will ask for the Clerk of Works' cooperation and assistance in carrying out these inspection duties. The contractor will be reminded of the obligation to provide the Clerk of Works with adequate facilities and access, together with information about site staff, equipment and site operations, and matters relating to site safety. The architect will explain procedures for verifying quality control, namely:

- certificates, vouchers, etc., as required
- samples of materials to be submitted
- samples of workmanship to be submitted prior to work commencing
- test procedures
- adequate measures for protection and storage
- visits to the workshops of suppliers and manufacturers.

3.1.5 Checklist: typical duties of a Clerk of Works

Typically, it will fall within the remit of the Clerk of Works to:

- examine work in the contractor's, subcontractors' and suppliers' workshops as necessary
- report according to the programme
- submit periodic reports as agreed
- keep a diary of events, register of drawings and file of instructions received, with relevant observations
- check drawings for errors, discrepancies and divergences, and notify the architect if any arise
- witness tests required by the contract or instructed by the architect
- monitor the application of specified techniques
- inform the architect of non-conforming work
- notify the architect immediately of problems arising and decisions needed
- confirm oral directions to the contractor in writing with a copy to the architect
- liaise with specialist Clerks of Works appointed
- check day work record sheets and record any wastage
- attend site meetings
- maintain as-built records and drawings
- record delays and the reasons for them
- take site photographs regularly and systematically, ensuring that they are date-endorsed and countersigned.

3.1.6 Actions beyond the scope of the Clerk of Works

Unless authorised by the architect or agent, the Clerk of Works will not:

- modify the design
- incur extra costs
- instruct the contractor about methods of working
- agree commitments with suppliers and subcontractors
- vary procedures specified in the contract
- issue instructions to the contractor unless authorised by the architect or employer's agent.

The Clerk of Works will not carry out instructions relating to the building contract from anyone but the person empowered under the contract to issue instructions.

3.1.7 Duties on site: quality matters

Quality control, as commonly practised on building sites, requires procedures to establish that work carried out and goods supplied conform to the standards specified in the building contract. Detailed records should provide identification, give the nature and dates of inspections, tests and approvals, and note the nature and extent of any non-conforming work found, together with details of corrective action. While this is the responsibility of the contractor, the Clerk of Works should keep independent records, which will be available to the client.

To the extent required by the contract, the contractor may be required to institute a quality management system. This will entail preparing a project quality plan showing how the contractor intends to meet the requirements. The project quality plan will normally give details of project organisation, reports on the performance of subcontractors, the work instructions or method statements to be followed, inspection and test plans listing the verification checks, and the planned dates for the project audits.

Effective Reporting for a Clerk of Works

4

4.1 EFFECTIVE REPORTING

4.1.1 A professional approach

In a sector that has become more reliant on fast, effective and efficient communication, it is important that the modern-day practitioner becomes more efficient in their reporting practices. Historically, Clerks of Works used carbon paperwork for issuing instructions and hard-copy diaries that would form part of their weekly reports. However, as courts and arbitration panels call for greater use of computerised evidence, it is vital for a Clerk of Works to become more familiar with all forms of digital reporting software. This is particularly relevant as we witness a surge in the introduction of innovative software packages, the wider exploration of SMART technology and the wider usage of Building Information Modelling (BIM). The next generation of practitioners will need to be accustomed to using IT and harnessing the power of these technologies to provide reporting methodologies relating to quality control matters.

The traditional duties of a Clerk of Works still apply — acting with ability, integrity and vigilance — but the way reporting is communicated, translated and transferred demands ever greater speed and efficiency.

4.1.2 Reporting proficiency

For Clerks of Works, the three principal vehicles for recording information remain. These are site directions, the site diary and periodic reports. However, the inclusion of more complex specialist reporting to support clients and fellow practitioners will invariably require closer interaction with project interventions and, in balancing increased workloads, may place greater reliance on photographic evidence or live streaming. Importantly, we are witnessing a move from previously agreed recording techniques to contract-specific agreed regimes of quality control methodology. Therefore the Clerk of Works must clarify the contract-specific reporting methodology before the start of the contract.

- Directions are still intended to pre-warn of the need for action, rectification or remediation works, but may be of no effect unless validated by the contract administrator.
- The site diary is for compiling a record of day-to-day events on site.
- Periodic reports give an assessment of progress on site and an intelligent forecast of requirements and developments going forward.

With the increased use of electronic recording and distribution, it is important to ensure that secondary back-up sources are used, such as cloud-based systems or USBs, ensuring that data is easily retrievable and password protected.

The basics remain constant, and the attention to accuracy, coherence, legibility and factual recording of events is the top consideration. Everything must be carefully recorded and attention paid to grammar, punctuation and spelling; ignoring these can lead to misunderstandings, errors and confusion.

4.1.3 Factual recording

The Clerk of Works should always scrutinise each item they have recorded to ensure trustworthiness, and ask the following questions.

- Does the recording capture the facts?
- Is it accurate?
- Is the recording unbiased?
- Does it read appropriately and make sense?
- Will the reader understand what has happened?
- Will the reader understand what needs to be done and the urgency of the matter?
- Would the use of pictorial evidence support the recording?
- If the recording was repeated as part of evidence in court, arbitration or adjudication, would it represent the facts in an appropriate manner and exclude any misunderstanding?

4.1.4 Site directions and non-conformity direction/instruction

With the introduction and wider use of bespoke contracts, and JCT, NEC and FIDIC forms of contract, the reference to the use of Clerk of Works direction may vary, and on certain forms of contract may be removed as a descriptive item/contract clause. However, some clients may still insist on the use of a Clerk of Works direction as an appropriate and relevant methodology of reporting early concerns relating to quality control issues. The terminology and reference may have been amended over the last decade, but the context and purpose of the direction remains unaltered.

A clear procedure for the issue and confirmation of site direction and/or non-conformity direction/instruction in accordance with the appropriate contract conditions (if applicable) must be established from the outset. In particular, in acknowledging the direction will not be valid until the architect/contract administrator confirms their acceptance of the direction by issuing an architect's instruction within 48 hours. It is imperative that all contractual parties understand the rationale, objective and purpose of the direction as a proactive early intervention.

Site directions must be coherent, and each direction should record the name of the contract, project name, date of issue, sequential serial number and any relevant supporting information that will ensure the contractor and relevant parties are clear on the message. Drawings should be referred to accurately, with their correct number, title and revision, and BIM drawings should be referred to through a code or reference from the software package. Actions, materials and techniques should be precisely described with times and locations as appropriate. The use of plain, straightforward language is important in communicating messages, and referencing supporting evidence, contract clauses, standards, illustrations, pictures or documentation will help the reader's understanding. Copies of all communications should be stored independently or backed up remotely.

4.1.5 Considerations before issue

The Clerk of Works/site inspector must verify with the architect/contract administrator before issuing any direction – in particular, where the direction:

- entails additional work
- has financial implications
- alters any of the contract particulars
- suggests approval of working methods that have not been agreed
- makes reference to use of new or alternative standards.

4.1.6 Frequency of entries

The frequency of reporting and the methodology of reporting techniques must be established at the outset with the architect/contract administrator/client. In most circumstances, keeping daily diary records is imperative, supported by periodic and weekly reporting. On larger projects this reporting regime can be as frequent as hourly, or zone specific, but in most cases the Clerk of Works' weekly report and monthly site meeting report forms the main framework of reporting and the first point of reference. With the use of modern IT systems this may include diary entries that are independently added to other practitioner reports or recorded on a main-frame reporting system.

Where modern application technology (apps) are used, it is important to ensure that diary entries have been correctly recorded. Some Clerks of Works take a secondary screenshot or picture of entries as their personal record.

4.1.7 The site diary

The site diary is for recording day-to-day events and accidents on site. The amount of detail to be included is a matter of judgement, but a Clerks of Works would do well to remember that the diary could be used as evidence in arbitration or litigation procedures. Therefore it is important to ensure that the entries are factual and report relevant and meaningful information in a format that is easily understood and accessible. The ownership and archiving of diaries and other supporting evidence should be agreed before the project begins.

As a minimum, entries made in a diary should include:

- verbal instructions or information from the architect/contract administrator
- information given by key consultants/practitioners
- day works and the reasons for day works

- weather conditions, especially extremes of temperature, high winds, rain and snow, etc.
- important construction events, such as concrete pouring and striking of formwork
- the start of important subcontract works; any problems of coordination or quality
- deliveries of goods and materials to site, including storage of materials
- movement of plant and equipment
- description of any tests witnessed, with names of personnel involved
- instances of poor workmanship, including records of if, and when, these matters were bought to the attention of the contractor
- any work or materials rejected, with reference to the relevant instructions, and how and when matters were rectified
- delays and reasons for these delays, either during an operation or between operations
- issue of drawings or information needed or requested, including date and time of issue, revisions and format of drawings, specifications or electronic data
- labour problems, overtime bans, strikes, difficulties surrounding labour numbers on site, etc.
- instances of inadequate site management or coordination
- any events, discussions or omissions, or any event that will have consequences relating to the contractor's/subcontractors' performance
- details of visitors to site
- specialist recordings, such as bespoke measurements, investigations or matters investigated on behalf of the client and/or other practitioners.

4.1.8 Effective use of the site diary

Whether the diary is hard copy or electronic, the following points outline a good practice methodology.

- The site diary should start from the date of appointment, or as soon as the project is occupied by the main contractor. In exceptional circumstances the client may want the Clerk of Work to take occupation during early ground investigation works. In this case, the diary should start from day one of operations.
- Items, interventions or events should be noted as soon as possible and, if records are recorded electronically, site notes and pictures should be uploaded at least once a day. In any event, daily records should be made on at least a daily basis.
- It is important to keep observations and records in the same sequence each day, as far as reasonably practicable.
- The diary recordings should be free of bias and should record evidence of facts.
- Where possible, the diary should include supporting evidence, pictures, emails, illustrations and extracts of conversations that will help the reader understand the context of a record.
- In all cases, a new diary/electronic log/electronic diary should be used for each project.

- If a separate notebook is used to assist in gathering data, this should be used to underpin evidence (if requested for legal proceedings, arbitration and/or adjudication).
- Where BIM or software data is used to exchange information, it is recommended that screenshots or a USB are used to retain evidence of data flow. In particular, where software compatibility issues are encountered, emails and copies of attachments recording delivery may prove invaluable.

4.1.9 Periodic reports

Historically, the Clerk of Works was required to provide the architect or client with periodic reports to record and assess on-site progress. Such reports were prepared on a weekly basis, and provided relevant parties with a 'snapshot' of progress and quality control matters. Today, as projects have increased in value and complexity, clients' expectations have changed and it is not uncommon to see a request for daily specialist reports to include a log of progress, or an events log for non-conformity items. For example, casting of specialist concrete works will require 100 per cent satisfaction before other sequential works can proceed.

Matters relating to opening up of works, repetition of works, claims, time extensions and delays can all have major impacts on critical path contract programmes, and clients/architects will need assurance that works are satisfactory before allowing further works to start. This puts additional pressure on the Clerk of Works, and appropriate contractual terms and conditions should all be included in the contract appointment. However, the use of digital recording devices and digital technology may help the Clerk of Works meet this more demanding reporting regime.

4.2 The role of technology

The increased use of IT systems, BIM and DIM technology and specialist software packages will revolutionise reporting/inspection techniques and will ensure that expectations of frequency of inspections and access to updated plans, specifications and drawings are fully fulfilled. Indeed, during the writing of this edition, technology is being piloted that will allow contract personnel to access the full range of project drawings via a visor built into a site helmet.

The introduction of iPads, smartphones and augmented reality (AR) technology will support and complement access to project data and ensure that, even on mega-projects, staff can access information within seconds. Traditionalists will always argue that an independent hard copy of the diary and site correspondence is recommended, but with cloud storage and big data collection archives, perhaps this can be left to personal choice or client expectations. In all cases, reporting should be adequately backed up, accessible and trustworthy. If unsure about the current guidelines on using technology, it is advisable to consult an IT specialist.

Chapter 4

4.2.1 BIM and Soft Landings

All Clerks of Works need to be aware of the government's drive towards BIM Level 2, and the Soft Landings principles now enshrined in legislation (BS 8536-1:2015).

Although the rationale behind this is for the architect, structural engineer, service consultants, etc. to fully coordinate their drawings towards federated and 3D modelling, there is a role to be played by the Clerk of Works in relation to the asset tagging requirement, and for the client's ongoing maintenance programming for the future life of the building.

BSRIA has produced a very useful guide – *Soft Landings Framework* (BSRIA BG 54/2014) – within which a section on 'Pit-stopping' explains this process further.

CDM Regulations 2015

5.1 CONSTRUCTION (DESIGN AND MANAGEMENT) REGULATIONS (CDM 2015)

The Construction (Design and Management) Regulations 2015 (CDM 2015) came into force in Great Britain on 6 April 2015. The Regulations apply to all construction work in Great Britain and the territorial sea, and set out what people involved in construction work need to do to protect themselves and anyone else from harm or ill-health.

CDM 2015 aims to improve health and safety in construction by helping dutyholders to:

- plan the work so the risks involved are managed from start to finish
- have in place the right people for the right job at the right time
- cooperate and coordinate work with others
- have the right information about risks and how they are being managed
- communicate this information effectively to those who need to know
- consult and engage with workers about the risks and how they are being managed.

CDM 2015 applies to any building, civil engineering or engineering construction work from concept to completion, and includes the following:

- construction, alteration, conversion, fitting out, commissioning, renovation, repair, upkeep, redecoration or other maintenance (including cleaning, which involves the use of water or an abrasive at high pressure, or the use of corrosive or toxic substances), de-commissioning, demolition or dismantling of a structure
- preparation for an intended structure, including site clearance, exploration, investigation (but not site survey) and excavation (but not pre-construction archaeological investigations), and the clearance or preparation of the site or structure for use or occupation at its conclusion
- assembly on site of prefabricated elements to form a structure or the disassembly on site of the prefabricated elements that, immediately before such disassembly, formed a structure
- removal of a structure, or of any product or waste resulting from demolition or dismantling of a structure, or from disassembly of prefabricated elements that immediately before such disassembly formed such a structure
- installation, commissioning, maintenance, repair or removal of mechanical, electrical, gas, compressed air, hydraulic, telecommunications, computer or similar services that are normally fixed within or to a structure.

For full details of the requirements of CDM 2015 refer to HSE document L153 – *Managing Health and Safety in Construction: Construction (Design and Management) Regulations 2015*.

5.2 THE ROLE OF THE CLERK OF WORKS

CDM 2015 makes no reference to the appointment or role of a Clerk of Works. However, under regulation 8, General Duties, there is a requirement to report anything that is likely to endanger dutyholders' own health or safety or that of others.

On many construction sites the principal contractor has a responsibility to those entering construction areas to demonstrate a minimum level of health and safety awareness, such as an in-date Construction Skills Certification Scheme (CSCS) card. It is advisable for a Clerk of Works to obtain a CSCS card in addition to passing the mandatory section on health and safety that is taken as part of the ICWCI membership and upgrade examinations.

This chapter provides a very brief overview of dutyholders' duties under CDM 2015. The aim is to provide awareness of this and other health and safety legislation in order to encourage regular, prompt and informed communication and cooperation with the design and construction teams.

5.3 CLIENT DUTIES

5.3.1 Who is the client?

A client is any person for whom a project is carried out. Where there is more than one client in relation to a project, one of them should agree to be treated as the client for the purposes of CDM 2015. This approach is to prevent the problems associated with a multi-headed client by avoiding conflicting instructions being given, and the confusion that can arise.

The main duties of the client are as follows.

- Make suitable arrangements for managing a project, making sure that:
  - o other competent dutyholders are appointed
  - o sufficient time and resources are allocated.
- Make sure that:
  - o relevant information is prepared and provided to other dutyholders
  - o the principal designer and principal contractor carry out their duties
  - o welfare facilities are provided.

Under CDM 2015 a domestic client's duties must be fulfilled by the sole or principal contractor unless there is a written agreement that the principal designer will fulfil those duties.

A domestic client is defined as someone who has work carried out on their own home or the home of a family member that is not done as part of a business, whether for profit or not.

5.3.2 Client duties in relation to managing the project

A client must make suitable arrangements for managing a project, including the allocation of sufficient time and other resources, and ensure that these arrangements are maintained and reviewed throughout the project. These arrangements are suitable if they ensure that the construction work can be carried out, so far as is reasonably practicable, without risks to the health or safety of any person affected by the project. The welfare facilities required by Schedule 2 of the Regulations are provided in respect of any person carrying out construction work.

A client must provide information in their possession, or which is reasonably obtainable by or on behalf of the client, that is relevant to the construction work and is of an appropriate level of detail and proportionate to the risks involved. This includes information about the project, planning and management of the project, and health and safety hazards, including design and construction hazards and how they will be addressed. It also includes providing information within any existing Health and Safety File as soon as is practicable to every designer and contractor appointed, or being considered for appointment, to the project.

Where there is more than one contractor, or if it is reasonably foreseeable that more than one contractor will be working on a project at any time, the client must appoint in writing a designer with control over the pre-construction phase as principal designer and a contractor as principal contractor. These appointments must be made as soon as is practicable, and in any event before the construction phase begins. If the client fails to appoint a principal designer, the client must fulfil the duties of the principal designer in regulations 11 and 12, which relate to the health and safety of the pre-construction phase and the Construction Phase Plan (CPP) and Health and Safety File. If the client fails to appoint a principal contractor, the client must fulfil the duties of the principal contractor in regulations 12 to 14, which relate to the health and safety of the construction phase of the project, the production of the CPP and Health and Safety File, and consulting and engaging with workers. In any case the client must ensure that these duties are complied with.

A client must ensure that:

- before the construction phase begins a CPP is drawn up by the contractor if there is only one contractor, or by the principal contractor when more than one contractor is required to complete the work, and;
- that the principal designer prepares a Health and Safety File for the project, which:
  - is appropriate to the characteristics of the project
  - contains information relating to the project that is likely to be needed to ensure the health and safety of any person during any subsequent project
  - is revised from time to time as appropriate to incorporate any relevant new information

o is kept available for inspection by any person who may need it to comply with the relevant legal requirements

o if a client disposes of the client's interest in the structure, is provided to the person who acquires the client's interest in the structure who will also be made aware of the nature and purpose of the file.

5.3.3 Notifiable projects

A project is notifiable if the work on a construction site is scheduled to last longer than 30 working days and have more than 20 workers working simultaneously at any point in the project, or exceeds 500 person days.

Where a project is notifiable, the client must give notice in writing to the Health and Safety Executive as soon as is practicable before the construction phase begins. The notice must contain the particulars specified in CDM 2015 Schedule 1, be clearly displayed in the construction site office in a comprehensible form where it can be read by any worker engaged in the construction work and, if necessary, be periodically updated.

Where a project includes construction work of a description for which the Office of Rail Regulation is the enforcing authority by virtue of regulation 3 of the Health and Safety (Enforcing Authority for Railways and Other Guided Transport Systems) Regulations 2006, the client must give notice to the Office of Rail Regulation instead of the Health and Safety Executive.

Where a project includes construction work on premises that are, or are on, either a GB nuclear site (within the meaning given in section 68 of the Energy Act 2013), an authorised defence site (within the meaning given in regulation 2(1) of the Health and Safety (Enforcing Authority) Regulations 1998), or a new nuclear build site (within the meaning given in regulation 2A of those Regulations), the client must give notice to the Office for Nuclear Regulation instead of the Health and Safety Executive.

5.4 GENERAL DUTIES

A person who is responsible for appointing a designer (including a principal designer) or contractor (including a principal contractor) to carry out work on a project must take reasonable steps to satisfy themselves that those people or organisations have the skills, knowledge and experience and, if they are an organisation, the organisational capability, necessary to fulfil the role that they are appointed to undertake, in a manner that secures the health and safety of any person affected by the project. Those people or organisations must in turn satisfy themselves that they have the skills, knowledge and experience and, if they are an organisation, the organisational capability necessary, and must not accept an appointment to a project unless they fulfil these conditions.

A person with a duty or function under CDM 2015 must cooperate with any other person working on or in relation to a project, at the same or an adjoining construction site, to the extent necessary to enable any person with a duty or function to fulfil that duty or function.

A person working on a project under the control of another must report to that person anything they are aware of in relation to the project that is likely to endanger their own health or safety or that of others.

Any person who is required by CDM 2015 to provide information or instruction must ensure the information or instruction is comprehensible and provided as soon as is practicable.

5.5 DUTIES OF DESIGNERS

A designer is any person (including a client, contractor or other person referred to in CDM 2015) who in the course or furtherance of a business prepares or modifies a design or arranges for, or instructs, any person under their control to do so, relating to a structure, or to a product or mechanical or electrical system intended for a particular structure. A person is also deemed to prepare a design where a design is prepared by a person under their control.

A designer must not start work in relation to a project unless satisfied that the client is aware of the duties owed by the client under CDM 2015.

When preparing or modifying a design the designer must take into account 'the general principles of prevention' (for further reference/definition of this term see HSE document L153 and any pre-construction information to eliminate, so far as is reasonably practicable, foreseeable risks to the health or safety of any person carrying out or liable to be affected by construction work, maintaining or cleaning a structure or using a structure designed as a workplace. If it is not possible to eliminate these risks, the designer must, so far as is reasonably practicable, take steps to reduce or, if that is not possible, control the risks through the subsequent design process. They should provide information about those risks to the principal designer and ensure that appropriate information is included in the Health and Safety File.

A designer must take all reasonable steps to provide sufficient information about the design, construction or maintenance of the structure to adequately assist the client, other designers and contractors to comply with their duties under CDM 2015.

Regulation 10 (CDM 2015) deals with people if established in Great Britain, or the client for the project who commissions construction, designs or modifications by a designer outside of Great Britain. This person or the client for the project must ensure that the duties summarised in this section are complied with.

This regulation does not apply to domestic clients.

5.6 DUTIES OF THE PRINCIPAL DESIGNER

The principal designer must plan, manage and monitor the pre-construction phase and coordinate matters relating to health and safety during this phase to ensure that, so far as is reasonably practicable, the project is carried out without risks to health or safety.

In fulfilling these duties, and in particular when design, technical and organisational aspects are being decided in order to plan the various items or stages of work, the principal designer must take into account the relevant content of any CPP and Health and Safety File. This work should be undertaken simultaneously or in succession, having estimated the period of time required to complete such work or work stages. The principal designer must identify and eliminate or control, so far as is reasonably practicable, foreseeable risks to the health or safety of any person who is:

- carrying out or liable to be affected by construction work
- maintaining or cleaning a structure
- using a structure designed as a workplace.

The principal designer must also ensure that all designers comply with their duties and that all people working in relation to the pre-construction phase cooperate with the client, the principal designer and each other.

The principal designer must assist the client in the provision of the pre-construction information required, and as far as it is within the principal designer's control, provide it promptly and in a convenient form, to every designer and contractor appointed or being considered for appointment to the project.

The principal designer must liaise with the principal contractor for the duration of the principal designer's appointment, and share with the principal contractor information relevant to the planning, management and monitoring of the construction phase, and the coordination of health and safety matters during this phase. The principal designer must also assist the principal contractor in preparing the CPP, by providing to the principal contractor all information the principal designer holds that is relevant to the construction phase plan including:

- pre-construction information obtained from the client
- any relevant information obtained from designers.

During the pre-construction phase, the principal designer must prepare a Health and Safety File appropriate to the characteristics of the project. This must contain information relating to the project that is likely to be needed during any subsequent project to ensure the health and safety of any person working on it. They must also ensure that it is appropriately reviewed, updated and revised to take account of the work and any changes that have occurred.

If the principal designer's appointment concludes before the end of the project, the principal designer must pass the Health and Safety File to the principal contractor. Where this is not the case, at the end of the project, the principal designer must pass the Health and Safety File to the client.

5.7 DUTIES OF CONTRACTORS

A contractor is any person, including a non-domestic client, who in the course or furtherance of a business, carries out, manages or controls construction work.

A contractor must not carry out construction work in relation to a project unless they are satisfied that the client is aware of their duties under CDM 2015.

A contractor must plan, manage and monitor construction work carried out either by the contractor or by workers under the contractor's control, to ensure that, so far as is reasonably practicable, it is carried out without risks to health and safety.

Where there is more than one contractor working on a project, a contractor must comply with any directions given by the principal designer or the principal contractor and the parts of the CPP that are relevant to that contractor's work on the project.

If there is only one contractor working on the project, the contractor must draw up a compliant CPP, or make arrangements for such a CPP to be drawn up, as soon as is practicable prior to setting up a construction site and take account of the general principles of prevention, when:

- design, technical and organisational aspects are being decided in order to plan the various items or stages of work that are to take place simultaneously or in succession
- estimating the period of time required to complete the work or work stages.

A contractor must not employ or appoint a person to work on a construction site unless that person has, or is in the process of obtaining, the necessary skills, knowledge, training and experience to carry out the tasks allocated to that person in a manner that secures the health and safety of any person working on the construction site.

A contractor must provide each worker under their control with appropriate supervision, instructions and information so that construction work can be carried out, so far as is reasonably practicable, without risks to health and safety. The information provided must include:

- a suitable site induction, where not already provided by the principal contractor
- the procedures to be followed in the event of serious and imminent danger to health and safety
- information on risks to health and safety:
  - o identified by the risk assessment under regulation 3 of the Management Regulations

o arising out of the conduct of another contractor's undertaking, and of which the contractor in control of the worker ought reasonably to be aware
● any other information necessary to enable the worker to comply with the relevant statutory provisions (e.g. instructions contained within the job method statement).

A contractor must not begin work on a construction site unless reasonable steps have been taken to prevent access by unauthorised people to that site.

A contractor must ensure, so far as is reasonably practicable, that the welfare requirements of Schedule 2 of HSE document L153 are complied with so far as they affect the contractor or any worker under that contractor's control.

5.8 DUTIES OF THE PRINCIPAL CONTRACTOR

The principal contractor must plan, manage and monitor the construction phase and coordinate matters relating to health and safety during the construction phase to ensure that, so far as is reasonably practicable, construction work is carried out without risk to health or safety.

In fulfilling these duties (in particular when design, technical and organisational aspects are being decided) the principal contractor must take into account the general principles of prevention.

The principal contractor must organise cooperation between contractors (including successive contractors on the same construction site) and coordinate implementation by the contractors of applicable legal requirements for health and safety. For the protection of workers and self-employed people, they must also ensure that employers:

● apply the general principles of prevention in a consistent manner, and in particular when complying with the general requirements for all construction sites listed in section 5.9 below and expanded on in Chapter 6
● where required, follow the CPP.

The principal contractor must ensure that:

● a suitable site induction is provided
● the necessary steps are taken to prevent access by unauthorised people to the construction site
● the welfare facilities listed in CDM 2015 Schedule 2 are provided throughout the construction phase.

The principal contractor must liaise with the principal designer for the duration of the principal designer's appointment and share with the principal designer information relevant to the planning, management and monitoring of the pre-construction phase and the coordination of health and safety matters during this phase.

During the pre-construction phase, and before setting up a construction site, the principal contractor must draw up a CPP or make arrangements for one to be drawn up. The CPP must set out the health and safety arrangements and site rules, taking account, where necessary, the industrial activities taking place on the construction site. Where applicable, it must include specific measures concerning work that falls within one or more of the categories set out in CDM 2015 Schedule 3.

Throughout the project the principal contractor must ensure that the CPP is appropriately reviewed, updated and revised so that it continues to be sufficient to ensure that construction work is carried out, so far as is reasonably practicable, without risk to health or safety. The principal contractor should provide the principal designer with any relevant information in their possession for inclusion in the Health and Safety File. Where the Health and Safety File is passed to the principal contractor on conclusion of the principal designer's appointment, the principal contractor must ensure that the Health and Safety File is appropriately reviewed, updated and revised from time to time to take account of the work and any changes that have occurred. At the end of the project the principal contractor must pass the Health and Safety File to the client.

5.8.1 Principal contractor's duties to consult and engage with workers

The principal contractor must:

- make and maintain arrangements that will enable the principal contractor and workers engaged in construction work to cooperate effectively in developing, promoting and verifying the effectiveness of measures to ensure the health, safety and welfare of the workers
- consult those workers or their representatives in good time on matters connected with the project that may affect their health, safety or welfare, in so far as they or their representatives have not been similarly consulted by their employer
- ensure that those workers or their representatives can inspect and take copies of any information that the principal contractor has, or that CDM 2015 requires to be provided to the principal contractor, which relates to the health, safety or welfare of workers at the site, except any information:
  - o the disclosure of which would be against the interests of national security
  - o which the principal contractor could not disclose without contravening a prohibition imposed by or under an enactment
  - o relating specifically to an individual, unless that individual has consented to its being disclosed
  - o the disclosure of which would, for reasons other than its effect on health, safety or welfare at work, cause substantial injury to the principal contractor's undertaking or, where the information was supplied to the principal contractor by another person, to the undertaking of that other person
  - o obtained by the principal contractor for the purpose of bringing, prosecuting or defending any legal proceedings.

5.9 GENERAL REQUIREMENTS FOR ALL CONSTRUCTION SITES

A construction site includes any place where construction work is being carried out or to which the workers have access, but does not include a workplace within the site that is set aside for purposes other than construction work.

A contractor carrying out construction work, or a domestic client who controls the work environment where any such construction work is carried out by an employee to provide safety measures and clear procedures, as noted below. The plan should include measures to:

- provide a safe place for construction work
- ensure good order and site security
- ensure the stability of structures affected by the works
- plan in writing and carry out demolition arrangements for the planning and execution of the works, or dismantling work, in a way that minimises risk
- ensure that, so far as is reasonably practicable, if used, explosives are stored, transported and used safely and securely and without risk of injury to any person
- ensure the safety of temporary works including excavations, caissons, cofferdams, falsework, formwork, scaffolds and gantries, arrangements for construction and removal, ensuring the stability of the works
- prevent risk from drowning
- control the risks from energy distribution installations
- ensure traffic routes are suitable and safe, and segregate vehicles and pedestrians
- ensure the safe use of vehicles and mobile plant
- prevent risk from fire, flooding or asphyxiation
- make suitable emergency arrangements
- provide sufficient fresh or purified air
- provide temperature and weather protection
- provide suitable and sufficient lighting.

These items are covered in more detail in Chapter 6: Health and Safety.

5.10 CONSTRUCTION PHASE PLAN

A CPP is prepared by a contractor or principal contractor and is a document that records the health and safety arrangements for the construction phase, site rules and, where relevant, specific measures concerning work that falls within one or more of the following categories:

- work that puts workers at risk of burial under earthfalls, engulfment in swampland or falling from a height, where the risk is particularly aggravated by the nature of the work or processes used or by the environment at the place of work or site

- work that puts workers at risk from chemical or biological substances constituting a danger to the safety or health of workers or involving a legal requirement for health monitoring
- work with ionising radiation requiring the designation of controlled or supervised areas under regulation 16 of the Ionising Radiations Regulations 1999
- work near high-voltage power lines
- work exposing workers to the risk of drowning
- work on wells, underground earthworks and tunnels
- work carried out by divers having a system of air supply
- work carried out by workers in caissons with a compressed air atmosphere
- work involving the use of explosives
- work involving the assembly or dismantling of heavy prefabricated components.

The plan must record the arrangements for managing the significant health and safety risks associated with the construction phase of a project. It is the basis for communicating these arrangements to all those involved in the construction phase, so it should be easy to understand and as simple as possible.

The information included must be relevant to the project, and have sufficient detail to clearly set out the arrangements, site rules and special measures needed to manage the construction phase. It should also be proportionate to the scale and complexity of the project and the risks involved.

The plan should include a description of the project, such as key dates and details of key members of the project team and how they will manage the work, including:

- the health and safety aims for the project
- liaison with the principal designer
- pre-construction information
- design information
- the site rules including personal protective equipment (PPE), parking, use of radios and mobile phones, smoking, restricted areas, hot works
- arrangements to ensure cooperation between project team members and co-ordination of their work, e.g. regular site meetings
- arrangements for involving workers
- site induction
- welfare facilities
- fire and emergency procedures
- the general principles of prevention of project risk (eliminate, reduce, isolate, control – known as 'ERIC')
- project changes
- securing the construction site
- demolition
- temporary works

- lifting operations (ensure all lifting operations are planned and supervised)
- permit to work systems
- Health and Safety File information
- monitoring the implementation and effectiveness of the plan — regular review, update and revision
- planning designated storage areas for materials and for waste.

Health and Safety

6.1 HEALTH AND SAFETY LEGISLATION

Health and safety is an integral part of any business, and the construction industry is no different from any other in this respect. The Clerk of Works has a responsibility principally to the client and architect, but also to liaise and work with the principal contractor and principal designer in respect of verifying quality of work carried out on site and materials used.

During site visits it is important that the Clerk of Works has sufficient knowledge of CDM 2015 and other health and safety legislation to be able to recognise shortfalls in a contractor's performance.

The Clerk of Works should be familiar with statutory regulations, guidance and advice published on the HSE website (www.hse.gov.uk) as relevant to the nature of the site and works being undertaken.

Here is a list of the primary health and safety legislation for construction:

- The Health and Safety at Work Act 1974
- The Construction (Design and Management) Regulations 2015
- The Management of Health and Safety at Work Regulations 1999
- The Workplace (Health, Safety and Welfare) Regulations 1992
- The Provision and Use of Work Equipment Regulations 1998
- The Lifting Operations and Lifting Equipment Regulations 1998
- The Manual Handling Operations Regulations 1992
- The Personal Protective Equipment at Work Regulations 1992 (as amended)
- The Regulatory Reform (Fire Safety) Order 2005
- The Control of Substances Hazardous to Health Regulations 2002 (COSHH)
- The Electricity at Work Regulations 1989
- The Confined Spaces Regulations 1997
- The Control of Noise at Work Regulations 2005
- The Health and Safety (First Aid) Regulations 1981
- The Reporting of Injuries, Diseases and Dangerous Occurrences Regulations 2013 (RIDDOR)
- The Work at Height Regulations 2005
- The Gas Safety (Installation and Use) Regulations 1998
- The Dangerous Substances and Explosive Atmospheres Regulations 2002
- The Health and Safety (Safety Signs and Safety Signals) Regulations 1996
- The Ionising Radiation Regulations 1999
- The Control of Pesticides Regulations 1986
- The Health and Safety Information for Employees Regulations 1989
- The Control of Lead at Work Regulations 2002
- The Control of Asbestos Regulations 2012

6.2 ARRANGEMENTS TO PLAN

The Clerk of Works should verify, with the principal contractor that the CPP currently reflects the activities taking place on site and that the site management team have in place adequate arrangements to monitor health and safety arrangements, performance and compliance with CDM 2015.

As the eyes and ears of the client on site, it is helpful for a Clerk of Works to verify that the principal contractor has procedures in place for:

- regular inspection and spot-checks of the CPP and health and safety procedures
- reporting to the architect, principal designer, single contractor, principal contractor and/or client, as applicable, any instances of non-compliance, and ensuring that corrective action is taken
- reporting on any incidents and delays due to work stoppage because of non-compliance.

6.3 FIRE SAFETY ON CONSTRUCTION SITES

Each year fire destroys many homes and businesses and takes lives, yet most of these fires could have been prevented. It is imperative to ensure that the significant site fire risks are assessed and recorded. A 'responsible person' should be appointed in the CPP with whom the Clerk of Works may liaise regularly with to verify that all reasonably foreseeable risks are accounted for.

The Health and Safety Executive publication HSG168 *Fire Safety in Construction* is available from the HSE website and gives guidance for clients, designers and those managing and carrying out construction work involving significant fire risks. Most HSE publications can be downloaded for free, or purchased from their website.

6.4 EMERGENCY PREPAREDNESS PLANNING

It is essential that adequate fire prevention, protection, detection, notification, suppression and escape measures are considered and provided, and that everyone on site knows what they are and what they need to do. All such arrangements must be clearly communicated at site induction, identified with relevant signage and regularly maintained in good order. Special consideration should be made to how the work will affect the emergency preparedness plan for areas adjoining, containing or contained by the construction site.

In addition to fire, many other incidents could result in an emergency with very little warning. Incidents caused by construction activities, security situations, extreme weather, medical issues and many other possible causes both on and off site need to be considered, and arrangements made for response to those that pose a significant risk.

Such Emergency Preparedness Plans (EPPs) must cover every aspect of the possible situation. Ultimately, the first objective is to prevent injury or loss of life, not just of those working on site, but members of the public as well. The other objective is to minimise potential losses to the business. It may be helpful that the Clerk of Works participates in the drafting of such plans, and that they are adequately tested. All people on site must be fully aware of their part of the plan.

Again, Clerks of Works may include in their own inspections and advise the architect, principal designer, single contractor, principal contractor and/or client, as applicable, of any instances of non-compliance.

6.5 HEALTH AND SAFETY SITE CHECKLISTS

It is important for a Clerk of Works to be aware of health and safety issues when visiting a construction site, to ensure that health and safety is being treated as an important element of the construction project.

The following checklists detail some of the elements that may be helpful to a Clerk of Works. The lists are not exhaustive and, depending on each project, specialist tools, equipment or processes may also be in use.

A wide range of additional information and detailed guidance is available from the HSE website.

6.5.1 Access and egress
- Is the site secure?
- Are arrangements in place to deal with visitors and workers new to the site?
- Can everyone reach their place of work safely?
- Are there safe roads with pedestrian routes segregated, suitable gangways, passageways, ladders and scaffolds?
- Are all walkways level and free from obstructions and trip hazards?
- Is edge protection provided at all leading edges to prevent falls?
- Are holes securely fenced or protected with clearly marked fixed covers?
- Is the site tidy and are materials stored safely?
- Is waste collected and disposed of properly?
- Are there chutes for waste to avoid materials being thrown from height?
- Are nails in timber removed or bent over?
- Is lighting provided for access routes and work tasks, particularly to dark or poorly lit areas?
- Are any necessary props or shores in place to provide temporary supports to structures or excavations?
- Are all health and safety notices clearly displayed, including emergency escape routes?

6.5.2 Cartridge-operated tools

- Are the maker's instructions being followed?
- Has the operator been properly trained?
- Is the operator aware of the dangers and able to deal with misfires?
- Does the operator wear eye protection?
- Does the operator wear ear protection?
- Is the gun cleaned regularly?
- Are the gun and cartridges kept in a secure place when not in use?

6.5.3 Compressed gases, e.g. liquefied petroleum gas (LPG), acetylene

- Are cylinders stored properly in lockable and ventilated cages with signage?
- Are all cylinders clearly identified?
- Are cylinders secured in an upright position, with the valve to the top?
- Is the cylinder valve fully closed when the cylinder is not in use?
- Are there 'hot work' procedures?
- Are cylinders in use for welfare facilities sited outside huts?

6.5.4 Cranes and lifting appliances

- Has a competent 'appointed person' (AP) been appointed to manage lifting operations?
- Is the crane inspected weekly, and thoroughly examined every 12 months by a competent person? Are the results of inspections recorded?
- Is there a test certificate?
- Is the driver trained, competent and over 18?
- Are the controls (levers, handles, switches, etc.) clearly marked?
- Does the driver and/or banksman find out the load's weight before trying to lift it?
- In the case of a jib crane with a capacity of more than one tonne, does it have an efficient automatic safe load indicator that is inspected weekly?
- In the case of a hydraulic excavator being used as a crane, is the maximum safe load clearly marked, and are hydraulic check valves fitted where required by the Certificate of Exemption?
- Is the crane on a firm and level base?
- Is there enough space for safe operation?
- Has the banksman or slinger been trained to give signals, and to attach loads correctly, and do they know the lifting limitations of the crane?
- If it can vary its operating radius, is the crane clearly marked with a Rated Capacity Indicator?
- Is the crane regularly maintained?
- Is the lifting gear in good condition, and has it been thoroughly examined (i.e. are slings and chains currently certified)?

6.5.5 Electricity

- Are all portable electric tools and equipment supplied at 110V, and have special measures been taken to protect them from damage and wet conditions?
- Are there any signs of damage to or interference with equipment, wires and cables?
- Are all connections to power points made by the correct electrical fittings?
- Are connections to plugs properly made so that the cable grip holds the cable firmly and prevents the conductors and earth wire from being pulled out?
- Are there 'permit to work' procedures where necessary to ensure safety?
- Are there any overhead electricity lines?
- Where anything might touch overhead lines or cause arcing (cranes, tipper lorries, scaffolding, etc.), has the electricity supply been turned off, or have other precautions been taken?

6.5.6 Emergencies

- Is there an emergency procedure in case of adverse weather, suspected un-exploded ordnance (UXO) discovery, bomb threats, etc.?
- Is the plan known by all on site, including visitors?
- Are muster points identified?
- Have the resources identified in the EPP been provided?
- Are these resources suitably protected and maintained?
- Is there a trained first-aider on site?
- Is there a suitable first-aid box?
- Are the first-aid box contents regularly checked?
- Is the accident book maintained and kept in a secure location?
- Is there an accident/incident reporting system in use?

6.5.7 Fire

- Has a Fire Plan for the site been produced?
- Does the site have the right number and type of fire extinguishers, as shown in the Fire Plan?
- Are there adequate escape routes?
- Are they kept clear?
- Do all workers know what to do in an emergency?
- Is there a warning system?
- Is it tested?
- Does it work, and can it be heard throughout the site?

6.5.8 Flammable liquids

- Is there a proper bunded storage area?
- Is the amount of flammable liquid on site kept to a minimum for the day's work?
- Is smoking prohibited?
- Are ignition sources kept away from flammable liquids?
- Are the correct safety containers used?

6.5.9 Other combustible material

- Is the amount of combustible material on site kept to a minimum?
- Are there proper waste bins?
- Is waste material removed regularly by an authorised operator?

6.5.10 Health

- Have all hazardous substances, e.g. asbestos, lead, solvents, etc., been identified and the associated risks assessed?
- Have workers been properly briefed on health hazards that might affect them?
- Can safer substances be substituted?
- Can exposure be eliminated, and if not can it be controlled at source rather than by using PPE?
- Is dust controlled?
- Have work activities that have the potential for hand-arm or whole-body vibration been assessed?
- If noisy work is to be undertaken, has a noise assessment been carried out?
- Are the workers nearby kept out of hazardous areas with warning signage displayed?
- In confined spaces, does the assessment consider atmosphere tests/alarms with a fresh-air supply provided if necessary?
- Are material safety data sheets available from suppliers?
- Have COSHH assessments been undertaken?
- Is skin protected from harmful radiation, including strong sunlight?

6.5.11 Excavations

- Have all underground services been investigated, located (with scanners and existing plans) and marked, and precautions taken to avoid them?
- Has an adequate supply of suitable timber, trench boxes, props or other groundwork-supporting material been delivered to site before any excavation work begins?
- Is a safe method in place for putting in and taking out the ground supports, i.e. one that does not rely on people entering an unsupported trench?
- If the sides of the excavation are sloped back or battered, is the angle of batter sufficient to prevent collapse?
- Is the excavation inspected by a competent person before each work shift and examined after any event that might have affected its strength or stability, such as after unexpected falls of materials?
- Is there safe access to the excavation, e.g. by a sufficiently long ladder, tied and at the correct angle?
- Are there barriers to prevent materials or people falling in?
- Are vehicles being kept away at a suitable distance from excavation edges to prevent collapse by vibration or additional ground loadings?
- If vehicles tip into the excavation, are properly secured stop blocks being used?

- Is the excavation affecting the stability of adjacent walls or neighbouring buildings?
- Is there a risk of the excavation flooding?
- Are materials, spoil or plant near the edge of the excavation likely to cause a collapse of the excavation?

6.5.12   Scaffolds
- If the scaffold has been designed and constructed for loading with materials, are the materials evenly distributed with sufficient space for walkways?
- Has the scaffold been erected by a competent person to the design requirements?
- Is there proper access to the scaffold platform, with self-closing gates to access points?
- Are all uprights properly founded and provided with base plates?
- Where necessary, are there timber sole plates, or is there some other way in which slipping or sinking can be prevented?
- Is the scaffold secured to the building, and are the ties strong enough?
- If any of the ties have been removed since the scaffold was erected, have additional ties been provided to replace them?
- Is the scaffold adequately braced to ensure stability?
- Are load-bearing fittings used where required?
- Are the working platforms fully boarded? Are the boards free from obvious defects, such as large knots/splits, and are they arranged to avoid tipping and tripping?
- Are the boards adequately fixed, particularly if high winds are forecast?
- Are there adequate guardrails and toe boards at every side from which a person could fall?
- Are there effective barriers or warning notices to stop people using an incomplete scaffold, e.g. one that is not fully boarded or is being altered?
- Does a competent person inspect the scaffold at least once a week, and always after bad weather or any other event that might have affected its strength or stability?
- Are loading bays properly gated and correctly used?
- Are the results of inspections recorded, including defects that were put right during the inspections, and are records signed by the person who carried out the inspection?

6.5.13   Falsework, formwork and temporary works coordinator
- Has a suitably skilled, knowledgeable and experienced temporary works coordinator or temporary works engineer been appointed?
- Does the coordination of temporary works include the principal designer in the review and liaison process in respect of the permanent design?
- Is there a temporary works register in place?
- Have all workers on, in or near temporary works areas been properly briefed?

- Is there a Temporary Works Method Statement, and does it deal with protecting workers from injury or ill-health?
- Has a falsework coordinator been appointed?
- Have the design and supports for shuttering and formwork been verified?
- Is it being erected safely from temporary working platforms or other safe means of access?
- Are the support bases and ground conditions adequate to take the loads?
- Are the props set out on a level base, plumb and cross-braced?
- Are the correct pins used in the props?
- Are the timbers in good condition?
- Is the falsework inspected by a competent person and checked against the agreed design before permission is given to pour concrete?

6.5.14 Ladders
- Has a risk assessment been carried out to justify the use of a ladder?
- Are ladders the right equipment for the job, or could a stair tower or other means of access be provided?
- Are Class 1 ladders in use, in good condition and properly positioned for access?
- Are they based on firm, level ground and at an angle of 1:4?
- Are ladders secured near the top, even if they will only be used for a short duration?
- If they cannot be tied at the top, are they adequately secured to prevent them from slipping?
- Do ladders extend at least 1.05m or five rungs above the step-off point?

6.5.15 Machinery and plant
- Are there any dangerous parts, e.g. exposed gears, chain drives, projecting shafts?
- Are dangerous parts adequately guarded?
- Are guards secured and in good repair?
- Is plant and machinery regularly inspected and maintained, with inspection records available?

6.5.16 Manual handling
- Have risk assessments been carried out?
- Can manual handling activities be avoided with the use of mechanical aids?
- If not, have the risks been assessed and reduced?
- Have those engaged in manual handling been properly trained?

6.5.17 Noise
- Do you need to raise your voice during normal conversation when 2m apart?
- Has a noise assessment been carried out?
- Is plant and machinery fitted with silencers/mufflers?

- Are hearing protection zones identified, with warning signage displayed?
- Do workers wear ear protection if they work in noisy surroundings?

6.5.18 Personal protective equipment/clothing
- Have risk assessments been carried out to determine what PPE is necessary?
- Is the protective equipment suitable for the employee?
- Has training in the use of the PPE taken place?
- Is the equipment provided suitable to protect site operatives from injury or ill-health?
- Is wet-weather gear provided for those who work in wet conditions?
- Do workers use and wear their PPE?
- Is PPE properly stored?
- Is replacement PPE available?

6.5.19 Method statements and risk assessments
- Are comprehensive site-specific method statements available?
- Have suitable risk assessments been carried out?
- Are risk assessments recorded?
- Have the general principles of prevention been properly considered?
- Is there evidence that site operatives have received training in the requirements of the method statements and risk assessments?

6.5.20 Risks to the public
- Have all risks to members of the public passing or adjacent to the site been identified, for example material falling from scaffolds, site plant and transport access and egress?
- Have precautions been implemented (e.g. scaffold fans/nets, banksmen, warning notices)?
- Is there adequate site-perimeter fencing to keep out the public, particularly children? Is the site secure during non-working periods?
- Are specific dangers on site made safe during non-working periods, e.g. are excavations and openings covered/fenced, materials safely stacked, plant immobilised and ladders removed or boarded?

6.5.21 Roof work
- Has the structural integrity of the roof been considered, particularly for refurbishment projects including those that are considered fragile, such as cement roofing sheets or those with roof lights?
- Has a plan of work been prepared as required by the Work at Height Regulations?
- Are roofing ladders or crawling boards available for use on roofs that slope more than 10 degrees?
- If not, is the roof pitch suitable for roof battens to provide a safe handhold and foothold, with a scaffold platform at eaves level to prevent falls from height?

- Are there guardrails or other temporary edge-protection measures to stop people or materials falling from roof edges?
- Are suitable guardrails, covers, etc. provided where people pass or work near fragile materials such as roof lights?
- Are fragile roof lights properly covered or provided with barriers to prevent falls?
- Are precautions taken to prevent tools, materials or debris falling onto others working under the roof?

6.5.22 Transport and mobile plant

- Are all vehicles and plant kept in good repair and regularly serviced?
- Do the steering, handbrakes and footbrakes all work properly?
- Are drivers and operators competent to operate the vehicle, and have they been trained, with proof of training held on file?
- Are all vehicles and plant being safely driven?
- Are vehicles securely loaded?
- Does the vehicle prohibit passengers?
- Are any tipping lorries being used, with a competent banksman overseeing operations?
- Is there a traffic management system in place to control on-site movements, with barriers and signage to separate vehicles from pedestrians?
- Are there separate routes and designated crossing points for pedestrians?
- Are vehicles being controlled by a competent and trained banksman following the safe system of work or traffic management plan?
- Do vehicles have audible alarms when turning or reversing, and are they fitted with rearview cameras?
- Are drip trays used for parked mobile plant?

6.5.23 Welfare

- Have suitable toilet and welfare facilities been provided in readily accessible places?
- Are rooms containing sanitary conveniences and washing facilities adequately heated, ventilated and lit?
- Are toilet, washing and welfare facilities being kept in a clean and orderly condition?
- Are separate rooms containing sanitary conveniences provided for men and women (except where each convenience is in a separate room, intended for the use of one person, the door of which is capable of being secured from the inside)?
- Are there clean wash basins, hot/warm water, soap and towels or other suitable means of drying provided in the immediate vicinity of all sanitary conveniences and changing room?
- Are wash basins large enough to wash hands, face and arms?
- Is there a room or area in which clothes can be changed, dried and securely stored?

- Is there a canteen, maintained at an appropriate temperature, where workers can take shelter, rest and have meals at tables with adequate chairs (with back supports) and the facility for boiling water and preparing meals?
- Is there an adequate supply of potable water and clean cups available?

6.5.24 Waste management

- Has a site management waste plan been developed?
- Have different categories of waste been identified, from low-risk to hazardous wastes?
- Has the progressive removal of waste materials been identified, to facilitate timely removal?
- Who is responsible for site waste removal?
- Are designated areas established for different categories of waste?
- Is a waste register established to identify the waste handling contractor and authorised tips?

6.6 CONCLUSION

Health and safety is a key element in any business, with construction considered a particularly high-risk industry. Getting it wrong could cost someone their life, or cause serious injury or result in a life-changing health problem.

Clerks of Works must keep themselves abreast of current developments in health and safety guidance and recognise bad work practices when visiting the site.

It is essential to observe what is happening and understand whether people are adhering to health and safety legislation requirements or are at risk of injury or ill-health. It is within the role of the Clerk of Works to highlight any significant shortfalls in good practice, and it is imperative to communicate with everyone involved in the project and to work with others in ensuring that construction sites are healthy and safe places to work.

In extreme circumstances, if a person is seen at immediate risk of injury, the Clerk of Works should ensure that they stop work and move to a place of safety. The site manager should be summoned and the Clerk of Works should explain the reasons for the intervention and obtain assurances that appropriate corrective measures will be put in place and records kept accordingly.

Trade Elements

7.1 DEMOLITION, ALTERATION AND RENOVATION

CONTRACT REQUIREMENTS

7.1.1 Ordnance benchmarks

Unless specified, the prior approval of the architect/employer's agent (A/EA) must be obtained when it is necessary to remove bricks, stones or other materials bearing benchmarks. The A/EA should report to Ordnance Survey any proposal to alter an Ordnance Survey benchmark, and the Clerk of Works should advise the A/EA when such action is completed, so that Ordnance Survey may be notified.

7.1.2 Preservation and protection

An early site meeting with the contractor should be arranged to ensure that all items or services to be preserved or protected in accordance with specification are correctly identified, marked and protected. Typically these might include the following.

- **Utility services:** the Clerk of Works should check that information from the providers is available, and agree any necessary points of disconnection. Service providers' conformity to requirements should be verified. The obtaining of wayleaves should be confirmed with the A/EA. It must be confirmed that the local authority has given approval.
- **Access:** the safety and adequacy of any maintained access for the occupying client must be verified.
- **Buildings and features:** adequate protection should be in place.
- **Trees, grass and plants:** the A/EA should agree the method of protection, if not specified. Normally, chestnut fencing maintained in good condition throughout the contract will be required, extending to enclose the full branch spread of the tree. No demolition fires are to be lit within 10m of the spread of trees to be preserved (if fires are permissible – the Environment Agency and many local authorities have restrictions and these should be verified with the A/EA).

Quality control

7.1.3 Demolition

All demolition work is subject to CDM 2015, and work should be carried out in accordance with the recommendations of BS 6187:2011 *Code of Practice for Full and Partial Demolition*, and the contractor must appoint a competent person to oversee the works. If the Clerk of Works considers that the contractor is carrying out any dangerous practice, which may threaten the safety of personnel, property or the public, the matter must be drawn at once to the attention of the contractor and the health and safety coordinator (the coordinator as required under the CDM Regulations 2007, or the planning supervisor under the CDM Regulations 1994). If satisfactory action is not taken, it should be brought to the notice of the A/EA, who may recommend, or action, an approach to the HSE Inspectorate. Actions should be recorded in the site diary.

7.1.4 Screens

The contractor must comply with any requirements in the specification or in relation to local bye-laws or planning conditions for the provision of screens to protect the public or private property.

7.1.5 Noise

The specification should be verified for any restrictions on weekend working, as construction sites that are close to residential areas can be a cause of noise nuisance. It is good public relations practice for a contractor to letter-drop all local residents to let them know what is going on.

7.1.6 Materials for reuse

The Clerk of Works should verify that operations relating to the recovery of materials to be saved for reuse, their cleaning as specified, and their proper storage and custody are being carried out correctly. Materials to be selected for reuse (e.g. bricks) should be approved.

7.1.7 Asbestos

In the event of the discovery of asbestos material it is the responsibility of the contractor to stop work and draw the matter to the attention of the A/EA and the health and safety coordinator. The Clerk of Works must advise the contractor, A/EA and the health and safety coordinator as soon as the presence of such material is discovered. Similarly, if the Clerk of Works considers that necessary action is not being taken, the contractor should be advised accordingly and the matter reported at once to the A/EA and the health and safety coordinator. Only approved licensed contractors can remove asbestos.

The major health and safety risk arises from asbestos dust, particularly from friable material, and damping down with sprayed water is recommended. All incidents involving asbestos must be recorded in the site diary. The HSE provides guidance on procedures and methods of working where a risk of exposure to asbestos has been identified. Work may not be restarted until formal clearance is given and a safe method of working has been detailed in the Health and Safety Plan, and implemented.

7.1.8 Contaminated land

Any contamination will normally have been identified by ground surveys, but if evidence of unexpected contamination is uncovered, the A/EA and health and safety coordinator must be informed at once. Samples should be taken and for analysis as instructed.

7.1.9 Temporary works

It is the responsibility of the Clerk of Works to receive and forward to the A/EA for approval the contractor's proposals for temporary weather protection and temporary supports to existing buildings or services undergoing alteration.

All temporary works and falsework should be designed and verified by the contractor. Calculations and drawings should then be forwarded to the A/EA/project structural engineer (SE) for appraisal and comment prior to any work starting.

The contractor will formulate a method of work as an instruction for the site staff to follow in conjunction with the falsework drawings. No loading should take place of any falsework until an authorised contractor's engineer has verified and approved its construction.

Regular checks should then be made to ensure that the falsework is not modified or stripped until the permanent structure is ready to be self-supporting.

Some categories of collapses of temporary works or falsework must be reported to HSE under the Reporting of Injuries, Diseases and Dangerous Occurrences Regulations 1995 (RIDDOR). Any problems must be notified immediately to the A/EA and the health and safety coordinator.

7.1.10 Adjoining buildings

It is the Clerk of Works' duty to receive and forward to the A/EA, or approve on the A/EA's behalf as directed, the contractor's proposals for the protection of adjoining buildings, including any information subject to the Party Wall Act 1996 (see also 3.1.3).

Statutory requirements and technical standards

Publications to which reference may be necessary include:

BS 5228-1:2009+A1:2014 *Code of Practice for Noise and Vibration Control on Construction and Open Sites: Noise*
BS 5228-2:2009+A1:2014 *Code of Practice for Noise and Vibration Control on Construction and Open Sites: Vibration*
BS 6187:2011 *Code of Practice for Full and Partial Demolition*
HSG 213 *Introduction to Asbestos Essentials*, 2001
HSG 227 *Comprehensive Guide to Managing Asbestos in Premises*, 2002
INDG188 *Asbestos Alert for Building Maintenance, Repair and Refurbishment Workers*
INDG223 *Short Guide to Managing Asbestos in Premises*, 2002
INDG289 *Working with Asbestos in Buildings*, 1999

7.2 GROUNDWORK

Contract requirements

7.2.1 A/EA's actions and approvals

The following is a checklist of items requiring action by the A/EA if specified, for which any delegation to the Clerk of Works in part or whole should be established, and to which the A/EA's attention should be drawn in good time as work progresses if this action or decision is required.

- **Surplus sub-soil:** to be spread where directed or removed from site by a licensed contractor.
- **Imported top soil:** samples to be obtained for approval by the A/EA before delivery.
- **Plants, trees and shrubs to be retained:** permission required to cut roots during excavation. Roots over 50mm should not be cut without the approval of the landscape supervising officer and the local authority tree preservation officer.
- **Turf for reuse:** permission to store for more than seven days, and approval of watering.
- **Herbicides:** approval of non-residual herbicide before spraying to clear surface vegetation.
- **Weather:** contractor to be advised of likely consequences of working in weather conditions specified as unsuitable. Events and conditions should be carefully recorded in the site diary and the A/EA consulted.
- **Supports to excavation:** approval of proposals, method statements and agreement of measurement if left in place or to be covered over.
- **Excess excavation:** permission to over-excavate, and instruction on backfilling.
- **Foundation bottoms:** approval to vary depth to suit ground conditions encountered.
- **Obstructions:** instructions and approvals for diversion of waterways, removal and filling of voids for old drains, manholes, foundations and other obstructions.
- **Permanent drains:** approval to use for disposal of water from excavations.
- **Antiquities:** permission for removal.
- **Explosives clearance:** on projects for MOD clients, it must be confirmed that the A/EA has obtained a clearance certificate and any necessary approval from the HSE Inspectorate and the local authority.

Quality control

7.2.2 Setting out

The contractor's setting out must be verified to ensure that it corresponds exactly with the drawings, and the contractor must verify and agree the original ground levels before excavation begins. Any disagreement must be reported to the A/EA, since it could have a bearing on quantities and payment.

Arrangements should be made to have temporary benchmarks established by the contractor checked against the nearest Ordnance Survey, or other validated benchmark, and recorded on the drawings.

7.2.3 Excavation and site clearance

Obstruction: details and dimensions of any large rock or other obstruction sufficient in size and difficulty to delay excavation and require special arrangements for its removal should be recorded for the information of the quantity surveyor.

Bottoms of excavations: the responsibility for carrying out stage inspections for compliance with Building Regulations may be delegated to the site inspector. The Clerk of Works should inspect the bottoms of excavations for foundations accordingly. The A/EA will make arrangements when necessary at the briefing meeting for the project civil engineer (CE) to inspect the bottoms of excavations with the Clerk of Works initially. Delegation of the duty should only be accepted when both parties are satisfied that the requirements are understood. Inspections are to be fully recorded in the site diary, and in situations where there is any doubt about the adequacy of a bearing surface the A/EA should be consulted.

Note: when the project is subject to statutory inspection by an approved inspector, the Clerk of Works should attend the inspection and record what transpires. If any problem arises the A/EA should be advised.

When foundation bottoms have been approved, concreting should proceed without delay. Any deterioration to the bottom or the sides before concreting begins is to be reported to the A/EA. It is the contractor's responsibility to provide protection against drying out or wetting of the excavation prior to concreting. If the contractor proposes, or the Clerk of Works sees the need for, any resealing or blinding of the bottom, or additional removal of failed side trenching, this should be referred to the A/EA for approval. Any such work carried out is to be recorded for the information of the quantity surveyor.

7.2.4 Materials for filling and method of compaction

Materials to be used for backfilling must be verified to ensure that they meet the specification. The thickness of layers and method of compaction must be confirmed to be as specified.

7.2.5 Piling: sheet, bored, driven, etc.

Piling is a highly specialised operation, and it is important for the Clerk of Works to establish with the A/EA what inspections and other tasks have to be carried out. The Clerk of Works will normally keep a daily progress record, showing on a site plan the location of piles, dates of boring or driving, sets obtained and results of pile testing. The Clerk of Works may also be required to verify setting out and tolerances, and to witness sets and tests. The project CE or SE will usually require records to be kept of particular conditions, for example changes of strata, appearance of water and obstructions encountered.

- **Advice:** normally advice should be obtained from the project CE under the A/EA's arrangements, but if specialist advice is required urgently, the client should be consulted (if employed directly), and they will advise on obtaining information on specific problems.
- **Safety:** contract preliminaries and/or the Health and Safety Plan should require the contractor to comply with the safety recommendations in BS 8004 and BS 8008 as appropriate. If the Clerk of Works is required to monitor piling operations, they must verify compliance. The contractor's attention must be drawn to any unsafe practice and, if it is not corrected, the matter brought to the attention of the A/EA and health and safety coordinator, who may recommend or action an approach to the HSE Inspector.
- **Noise:** piling is often the noisiest operation on site, and particular limitations may be included in the specification. Special briefing should be provided on any Clerk of Works' duties in this respect, which may range from simply recording periods of piling operation to carrying out noise measurements, and operations must comply with the Noise at Work Regulations 2005. BS 5228-1:2009+A1:2014 and BS 5228-2:2009+A1:2014 provide further information. Good public relations can often help to offset the noise nuisance and the Clerk of Works may be called upon to play a part in this.

7.2.6 Diaphragm walls

This is a specialised form of construction for which particular briefing from the A/EA/SE should be obtained in respect of Clerk of Works' duties. These may include delegated inspections for leakage. BS 8102 relating to waterproofing of structures below ground provides detailed information. Supports to the wall may be provided in temporary or permanent construction, using strutting or ground anchors. Instruction should be obtained from the A/EA or SE on any duties of load testing of supports and de-stressing of anchors.

7.2.7 Underpinning

Particular duties of the A/EA, for which any delegation should be agreed, or for which notification in good time is required, are as detailed below.

- **Sequence of operations:** this and any special procedural requirements are to be agreed with the contractor. Detailed record drawings showing the sequence of operation and curing times should be maintained during this element of the works. Reasons relating to any delays should also be recorded.
- **Simultaneous underpinning:** the A/EA is to approve simultaneous work on adjacent sections at less than a specified limit apart.

- **Supports:** A/EA's approval is required to leave any supports in place when backfilling.

Statutory requirements and technical standards

Publications to which reference may be necessary include:

BS 4428:1989 *Code of Practice for General Landscape Operations (Excluding Hard Surfaces)*
BS 5228-1:2009+A1:2014 *Code of Practice for Noise and Vibration Control on Construction and Open Sites: Noise*
BS 5228-2:2009+A1:2014 *Code of Practice for Noise and Vibration Control on Construction and Open Sites: Vibration*
BS 8000-1:1989 *Workmanship on Building Sites. Code of Practice for Excavation and Filling*
BS 8004:2015 *Code of Practice for Foundations*
BS 8102:1990 *Code of Practice for Protection Against Water of Structures Below Ground*

7.3 IN SITU CONCRETE/LARGE PRECAST CONCRETE

Contract requirements

7.3.1 A/EA's approval

Approvals required for which the Clerk of Works should establish delegation from the A/EA include the following.

- **Details, drawings and calculations:** these will normally be required only where temporary works are large and complex, and in such cases are likely to be dealt with by the A/EA or project CE. The Clerk of Works may, however, be called upon to assist.
- **Formed finishes:** arrangements for control sampling are to be agreed and examples or photographs of required finishes for inspection purposes obtained from the A/EA or contractor as specified. In the case of superfine and other special finishes, the A/EA's approval is required for details such as panel size and arrangement, and use and location of cover spacers and of battens.
- **Excavation faces:** use of these as formwork is not advised unless they are rigid and will stand up to the concrete being placed and the vibration of a vibrating poker and are able to support workers along the edges. It is essential that some form of sheeting barrier be placed between the cut excavation face and any reinforcement.
- **Surface retarders:** the use of these is to be strictly in accordance with the manufacturer's recommendation and the permission of the design CE.
- **Striking formwork:** the minimum periods to be observed before striking different elements of the formwork should be specified, and may vary with surface

temperature of the concrete and its strength. It is the responsibility of the Clerk of Works to verify that specifications are correctly observed. Construction loading, which can be significant, must be taken into account. Re-propping is not permitted without the A/EA's approval, and in any case of doubt the A/EA must be consulted.

7.3.2 Quality assurance

The quality assurance (QA) system to be implemented must be agreed with the A/EA, and instructions obtained. In this workgroup, reference is made to the BSI Registered Firms Scheme and CARES (the UK certification body for reinforcing steels).

7.3.3 Quality control

Concrete is a complex product, comprising many materials, which is required not only to comply with specified standards but also the variable requirements of workmanship. Supervision of concrete work depends on a full understanding of the code to be followed and the designer's intention. Where the responsibilities are divided between the project CE, the A/EA and the site inspector, as is usually the case, it is essential that a clear brief from the A/EA on the Clerk of Works' duties is obtained, and that any especially important points to watch are highlighted.

7.3.4 Setting out

The setting out must be verified as directed by the A/EA. This will normally comprise verification of primary setting out of main features such as stairwells and lift wells, random checks on grid, verticality and cambers, section sizes in formwork for beams, and interface with beams and columns.

7.3.5 Formwork

The formwork must be thoroughly clean, undamaged and free from debris. Verification for this are best done when the reinforcement is being inspected, immediately prior to pouring concrete. All inserts, holes, channels, 'blow-holes' for debris removal, embedded parts and services must be confirmed as being in place, and checks made to see that formwork has not been damaged by loads imposed on spacer blocks. If the contractor proposes to pour the concrete in lifts, it must be established that this is acceptable to the A/EA or project CE.

Cutting away reinforced concrete for service runs is not permitted without the A/EA's prior approval. The Clerk of Works must therefore verify with the M&E principal site manager (PSM) representative that provision has been made for service runs by boxing or another specified method.

7.3.6 Concrete mix and materials

The contractor should submit for the A/EA's approval proposals for meeting the contract specification for concrete, covering sources of materials, suppliers, mixes, chlorides, alkalis and other aspects of concrete quality. Concreting may not be started prior to approval of submitted proposed 'design mixes'. All copies should be verified and forwarded to the A/EA. Three copies should be returned to the Clerk of Works on approval: two for the contractor (one for the ready-mix supplier) and one for retention. Where pump mixes are to be used, the specification design will need approval from the SE.

Ready-mix: it is very important that where mix designs are based on a ready-mixed supply, the contractor has provided the ready-mix supplier with a copy of all the relevant specification clauses. There should be a specification clause detailing this requirement and compliance must be verified. The A/EA will arrange for any inspection of the ready-mix plant.

Material orders: early orders by the contractor on suppliers of materials or ready-mixed concrete should be verified for full conformity to the specification. This check may be omitted when the Clerk of Works has full confidence in the correctness of the orders supplied, but should be reimposed with any change in responsible staff.

Certification: all certificates, calculations and other documents relating to supply of materials as required by specification must be obtained and submitted to the A/EA for verification. A record is to be kept of these submissions.

Delivery and storage: the Clerk of Works must verify that delivery and storage provisions are as specified, with the aim of ensuring that materials are kept clean, uncontaminated and weather protected, and that they are used in order of delivery.

A/EA's approval: any delegations must be agreed and the matters requiring the A/EA's attention should be highlighted in good time as work progresses. The following is a checklist of items relating to materials:

- **admixtures:** admixtures may only be used subject to prior approval by the A/EA; applications should be submitted to the A/EA quickly to avoid delay to the works
- **source of aggregates:** to be approved
- **colour:** where specified, samples of aggregate to be approved
- **water:** any source other than mains supply to be approved.

7.3.7 Workmanship

Trial mixes: the A/EA may require trial mixes in connection with the approval of mix design proposals. The Clerk of Works should supervise the preparation of trial mixes as directed by the A/EA and keep records. The A/EA's attention should be drawn to any change in constituent materials during the progress of the works that might justify a repeat of the trial mix.

Site concrete production: batch mixing is required for site production of concrete. Satisfactory conditions for mixing plant, which should conform to regulations, must be confirmed. During production, the accuracy of the weighing machines should be checked each morning, most conveniently by weighing unopened bags of cement. Batching by weight is required for all solid constituents, except that batching by volume of aggregates is permitted for certain mixes as specified. The regular use of any mixer bigger than necessary for the task should be discouraged since this may lead to error or misuse. Cement should be added by measured number of whole bags, or by weighing on a machine used only for weighing cement. Where pigments or admixtures are to be added, methods of doing so must be confirmed as satisfactory. Cement should be used in date order, from the date of manufacture on the bag.

On sites with large-scale production, the A/EA will lay down a schedule for periodic verification of the calibration of all measuring devices.

Safety: dry cement mixed with water releases an alkali, which can be harmful. Contact between the skin and fresh concrete or mortar should be avoided by wearing suitable clothing. Concrete or mortar adhering to the skin should be washed off with clean cold water as quickly as possible, or skin irritation may result.

Workability: workability must be appropriate to the casting operation. The Clerk of Works should witness the contractor's slump or compaction factor tests, and carry out their own tests if there is any suspicion that the workability is unsatisfactory.

The contractor and A/EA must be informed if the Clerk of Works is not satisfied, and a record made in the site diary.

Placing concrete: conformity with specification must be verified, and the following points requiring site approval noted, which should be agreed as necessary with the A/EA or SE in advance of the work:

- **cold weather:** approval to place concrete at a temperature below the limit in specification, and of methods to be used to maintain temperature of concrete during and after placing, e.g. thermal matting
- **wet weather:** approval to start concreting in the open when it is raining heavily, and of the method of protecting finished surfaces from heavy rain.

It should be remembered that concrete should not be dropped into the shuttering from such a height that could cause segregation of the mix, and time periods between pours should be strictly adhered to. Suitable stop-ends and day-joints will need approval from the A/EA and SE prior to pouring.

7.3.8 Curing concrete

Method: the Clerk of Works should discuss with the contractor the proposed methods of curing, and verify that all the necessary materials are readily at hand before pouring. The A/EA should be consulted in the case of any doubt arising. Where specified, the A/EA's prior approval for the method proposed should be obtained. Weather conditions should be taken into account and the need to vary curing times considered, as specified.

It is the responsibility of the Clerk of Works to verify that:

- curing starts at the proper time
- any curing compound is sprayed on uniformly
- polythene membrane sheeting is sealed at edges and joints, has no tears and is weighted down throughout curing
- hessian or similar material is regularly watered and kept wet until the end of curing
- formwork is kept in place if no other curing is provided instead
- surfaces are protected from solar gain in hot weather
- curing compound is fully degraded before the application of any finishes and by the end of construction.

7.3.9 Checking cover to reinforcement

The A/EA and SE must be notified in good time before programmed work of covering, cladding or finishing of concreting, ready for verification. The faces to be verified are those specified, and any others as instructed by the A/EA. The A/EA should be advised of any additional areas where a check is considered advisable. Verification should be carried out where so directed by the A/EA, with the assistance of the contractor, and the A/EA notified if the cover is found not to meet the requirements of the drawings or the specification.

7.3.10 Watertight construction

Workmanship: watertight construction requires a high standard of workmanship, and the Clerk of Works should monitor compliance with the specification very closely, verified with the contractor in advance whether form ties are to be used and the acceptability of the ties. The A/EA should be notified of any necessary check on tie stress. Kickers must be formed at the same time as the slab, not separately, and any pipe penetrations must incorporate puddle flanges to form water barriers.

Inspection: the A/EA must receive due notification when joint inspections with the contractor are to be carried out as specified (i.e. after completion of slab-over and after the main frame of the building is complete), or the Clerk of Works may inspect on the A/EA's behalf as directed and report any signs of leakage.

Grouting: if grouting has to be carried out, the A/EA's approval of the proposed method should be obtained, and the A/EA notified when work is ready for inspection, or the Clerk of Works may inspect on the A/EA's behalf as directed. Socket holes etc. may not be made good until the grouting method is accepted as satisfactory.

7.3.11 Testing of concrete

Refer to BS 8500-1:2015+A1:2016, BS 8500-2:2015+A1:2016 and BS EN 206:2013+A1:2016.

Test cubes: the proper casting of test cubes by the contractor must be witnessed and verified, in accordance with the specification. Cube moulds should be calibrated to BS 1881 and QSRMC requirements. The A/EA's approval for the contractor's proposed independent laboratory must be obtained. Some concrete delivery firms will certificate instead of offering a cube-testing service.

Independent tests: any requirements and delegation to the Clerk of Works to take cubes for independent testing should be discussed with the A/EA, e.g. for sensitive elements of the structure or where there has been any problem, and samples despatched to the laboratory as directed by the A/EA.

Workability: random sampling should be carried out at the specified rate per m^3. The Clerk of Works should witness and record measurements of workability in accordance with the approved method.

Records: a record of concrete test cubes taken and a record of the results of cube tests must be kept. A record of the locations from which cubes were taken should be made on appropriate drawings and cross-referred to the relevant form.

7.3.12 Steel reinforcement

Quality assurance: reinforcing steel is normally to be supplied by a firm belonging to the CARES scheme (the UK certification body for reinforcing steels), and is also to be cut and bent by a CARES licensee. Certificates should be obtained from the contractor for suppliers showing the CARES certificate of approval number and forwarded to the A/EA.

Materials: the Clerk of Works should verify that steel bars and fabric bear the CARES marking (see CARES leaflet; www.ukcares.com) and that delivery notes bear the CARES certificate number.

Workmanship: the work of a CARES licensee should not require close inspection, but it remains essential to carry out sample checks on bar diameters and other dimensions, spacing, and positioning of reinforcement, especially in critical areas. If there are any doubts about the standard of work, the A/EA should be consulted.

Site work carried out by anyone other than a CARES licensee should be confined to minor adjustments and cutting of random lengths of secondary reinforcement made on site. The A/EA should agree on the delegation to approve cold bending machines and the conditions under which hot bending and re-bending or straightening after fixing may be approved. No site heating or welding of reinforcement should be permitted without the approval of the A/EA. The use of couplers may be acceptable, but the A/EA should be consulted.

Spacers and chairs: any delegation of approval from the A/EA for material and type of spacers and sizes of chairs should be confirmed. Details are normally the responsibility of the contractor, but in special circumstances, such as deep slabs or areas of congested or complex reinforcement, details may be included in drawings or bending schedules. The Clerk of Works should verify adequacy and, when wood-wool permanent formwork is used, verify that supports are sufficient to avoid indentation. The A/EA should be informed in the event of any dissatisfaction. During the pouring of concrete, the Clerk of Works should verify that spacers are stable and not displaced, and that reinforcement maintains its specified clearance from the shuttering.

Discrepancies: it is the responsibility of the contractor to draw attention to any incompatibility with the drawings. Any discrepancy that cannot be simply resolved should be reported to the A/EA for instruction.

Verification tests: these should not be required for materials supplied under a CARES licence, but if any doubt arises about the quality of material, it should be reported without delay to the A/EA for instructions. The location(s) of suspect material must be recorded. If the A/EA directs that a verification test be carried out, the Clerk of Works should identify the suspect material from which test pieces are to be cut. The results of tests should be forwarded to the A/EA.

7.3.13 Precast and composite concrete decking

Components for such work will be largely the subject of off-site or other specialist inspection and testing. The Clerk of Works should obtain briefing from the A/EA on any particular duties, but they will normally include verification of the condition on delivery to site and subsequent checks in the course of construction, for example during propping, erection and placing of in situ topping.

7.3.14 Suspending and rejecting concreting work

The contractor should be warned of the advisability of suspending work if there are compelling reasons. The A/EA must be consulted in all such cases.

Rejection of completed work will normally require consultation with the project SE, who will wish to verify the site diary records of casting. Typical examples of situations where rejection may be called for include:

- failure of cube tests to meet specified limits — at least four test results per pour are required, and decisions rest with the A/EA, who may order loading tests where cube results have shown failures
- movement of formwork or reinforcement during pour
- unsatisfactory finish, including any 'honeycombing'
- failure of steel validation test
- excessive cracking

- lack of proper curing
- failure to protect against frost.

7.3.15 Building Regulations

The Clerk of Works must monitor to ensure that the contractor is acting in compliance with the Building Regulations and that inspection is carried out prior to covering up of foundations, or oversite concrete with other oversite material. Where the project is subject to statutory inspection by an approved inspector, the Clerk of Works should attend the inspection and record what transpires. The A/EA should be informed of any problem that arises.

Statutory requirements and technical standards

Publications to which reference may be necessary include:

Building Regulations 2000
BS 8000-2.1:1990 *Workmanship on Building Sites. Code of Practice for Concrete Work. Mixing and Transporting Concrete*
BS 8000-2.2:1990 *Workmanship on Building sites. Code of Practice for Concrete Work. Sitework with In Situ and Precast Concrete*
BS 8110-1:1997 *Structural Use of Concrete. Code of Practice for Design and Construction*
BS 8500-1:2015+A1:2016 *Concrete. Complementary British Standard to BS EN 206. Method of Specifying and Guidance for the Specifier*
BS 8500-2:2015+A1:2016 *Concrete. Complementary British Standard to BS EN 206. Specification for Constituent Materials and Concrete*
BS EN 206:2013+A1:2016 *Concrete. Specification, Performance, Production and Conformity*

7.4 MASONRY

Contract requirements

7.4.1 A/EA's actions and approvals

It is the responsibility of the Clerk of Works to go through the specification and identify all items in material and workmanship sections that require an action or approval by the A/EA (normally but not always specified as A/EA's responsibility). These should be brought to the notice of the A/EA in good time as the works progress, and any delegations, in part or whole, established. Typical examples are listed below.

Samples: where approval of samples or sample panels is specified, the Clerk of Works should ensure that these are submitted or that readiness is notified to the A/EA before the contractor confirms an order. An approved sample should be obtained for retention on site to match against deliveries. Random samples should be taken as specified from initial deliveries and approved or referred to the A/EA as delegated. If there are any doubts about compliance the A/EA should be consulted.

Tolerances: normally to BS EN 771-1:2011+A1:2015 and BS EN 771-2:2011+A1:2015 for bricks and BS EN 771-3:2011+A1:2015 and BS EN 772-2:1998 for blocks. Any relaxation of specified tolerances must be agreed with the A/EA.

Tipping of loads: tipping of loads of masonry units should not be allowed unless this has been authorised by the A/EA.

Carved stone: the A/EA will normally approve the selection of stone for carving and will decide whether profiles will be carved before or after the stone has been built in.

Finishes and facework: any special requirements should be identified and the A/EA's instructions and any delegation of approvals obtained.

Overhand work: normally permitted only where the bricklayer cannot gain access to a normal working position.

Admixtures: the specification may permit plasticisers to BS EN 934-3:2009+A1:2012 to be used in mortars, but any proposal to use other admixtures should be referred to the A/EA for a decision. The use of calcium chloride by itself or in admixtures is not permitted.

Alternative mixes: any proposal for an alternative mortar mix should be referred to the A/EA for a decision. The specification may also permit the use of factory-produced pre-mixed and retarded mortars, or batched mortars produced and delivered to site. If site mixes are agreed, the method of gauging for mortars should also be established. In all cases the manufacturer's instructions are to be adhered to precisely.

Frost damage: the contractor is required to take down and rebuild any work damaged by frost. If any doubts arise the A/EA should be consulted.

Broken bonding: if the masonry bond is not specified, the A/EA's instructions should be obtained. Where bonding is specified, the A/EA's instructions on the location and arrangement of any broken bond in faced work must be sought. On corners, 'toothing out' should not be allowed unless specifically required. 'Racking back' is best practice on corner work.

Rate of progress: any proposal to raise walls at a pace above that specified should be referred to the A/EA for approval or otherwise. It is good practice to bring the walls up together if possible, and either leaf should not be raised more than 1.5m high before the second leaf is added.

Wall ties, cavity trays and weep-holes, etc.: the A/EA should be consulted for types of wall tie permitted, and whether fixed, sloping or level. Refer to the manufacturer's instructions. Fire stopping and barriers, along with cavity insulation should be confirmed as appropriate to the design, and both fire and cavity closers are to be mechanically fixed and not left fitted loose.

7.4.2 Quality assurance

The specification should be verified for applicable QA schemes, and the A/EA's instructions obtained on any duties arising. Agrément certification is used in this workgroup.

Quality control: it is one of the duties of the Clerk of Works to verify conformity with the specifications and drawings. Where materials are specified to British Standards or Agrément certificates, the appropriate markings must be verified and any additional or particular requirements that may require certification or test should be noted.

7.4.3 Setting out

All setting out of masonry must be verified in relation to the established centre lines of the building and the given building line. All dimensional verifications must be made with a steel tape. Additional checks should be carried out as the work rises. The Clerk of Works should be familiar with the bonds specified, ensure that the setting out of the first course is accurate, and that the work is related to established datum levels. A datum at finished ground floor level is a convenient point. The use of storey rods for setting out storey heights from datum and showing the tops and bottoms of openings, course levels, etc. should be encouraged.

7.4.4 Weather protection

The adequacy of the protection of materials from frost and rain should be confirmed, as should the provision of insulated covering to new masonry walling where specified for low temperatures. Newly erected masonry should be top covered during rain to prevent saturation and future efflorescence and mortar being washed out of joints.

7.4.5 Certification

In addition to conformity to relevant British Standards, bricks and blocks may be required to meet particular specifications of type, colour, strength, durability, water absorption, manufacturing control level, etc. The Clerk of Works should verify that the contractor's orders properly cover these requirements, and verify the delivery documentation for evidence of compliance. The A/EA's instructions should be obtained on acceptable certification. Some suppliers may meet QA standards, in which case their certificate of compliance will normally be acceptable (e.g. BSI Kitemarked and Agrément-certificated products). In other cases the A/EA may require independent testing.

7.4.6 Testing

Testing is normally required only for structural masonry. Initial testing of mortars to be used is to be carried out by the contractor at least six weeks before commencing masonry work. The Clerk of Works should witness preparation of test specimens and record details of mix and ball penetration tests. Comprehensive strength test certificates received should be submitted to the A/EA for approval to start masonry work. The A/EA's instructions should be obtained on further testing to be carried out as the work progresses.

Bricks and blocks: independent testing should be arranged as directed by the A/EA. The following are normally tested:

- comprehensive strength of load-bearing bricks and blocks
- water absorption of facing bricks and those to be used below the damp-proof course
- thermal conductivity of dense or lightweight aggregate or aerated concrete blocks for load-bearing walls and partitions.

7.4.7 Scaffolding

If it is not specified, the A/EA's decision should be obtained as soon as possible on whether putlog or independent scaffolding is to be used.

7.4.8 Manufacturer's instructions

The Clerk of Works should obtain copies of the manufacturer's instructions and Agrément certificates where applicable, and verify that they are followed.

7.4.9 Weighing plant

If weighing machines are employed in site mortar-mixing plant, the Clerk of Works should carry out or witness regular verification of their accuracy. The maker's instructions should be followed on this. An unopened bag of cement makes a convenient weight for checking.

7.4.10 Conditioning bricks and blocks

Where a conditioning period is specified, the date of supply should be noted and checks carried out to ensure the material is not used before the period expires.

7.4.11 Damp-proof course (DPC)

In addition to the specified workmanship, attention should be paid to the following points.

- Rolls of bitumastic DPC should be verified to ensure that they do not crack when unrolled in cold weather.

- Where the DPC is stepped, the DPC may not be laid less than 150mm above the finished ground level.
- The edge of the DPC must not be bridged by mortar. It should be flush or slightly projecting. If used as a drip, it should extend 5mm.
- If cavity walling has to be cleaned out, it must be established that the method employed will not damage the DPC or any insulant.
- The connection and sealing between the horizontal wall DPC and the damp-proof membrane to a floor must be verified.
- Vertical DPCs should be left proud of brickwork around doorways, etc., and extend beyond horizontal DPCs for lapping.
- The Clerk of Works must be vigilant where the specification provides for forming bridges with DPC material. If the contractor proposes the use of pre-formed cavity trays, these should be Agrément-certificated.

7.4.12 Services

Where services penetrate an exterior wall below ground, in addition to sleeving and sealing against passage of gas it may be necessary to provide protection against ingress of groundwater. The A/EA's instructions should be obtained.

Statutory requirements and technical standards

Publications to which reference may be necessary include:

BS 3921:1985 *Specification for Clay Bricks has been replaced by the following*:
BS EN 772-3:1998 *Methods of Test for Masonry Units. Determination of Net Volume and Percentage of Voids of Clay Masonry Units by Hydrostatic Weighing*
BS EN 772-7:1998 *Methods of Test for Masonry Units. Determination of Water Absorption of Clay Masonry Damp Proof Course Units by Boiling in Water*
BS EN 771-1:2011+A1:2015 *Specification for Masonry Units. Clay Masonry Units*
BS 5628-3:2005 *Code of Practice for the Use of Masonry. Materials and Components, Design and Workmanship has been replaced by the following:*
PD 6697:2010 *Recommendations for the Design of Masonry Structures to BS EN 1996-1-1 and BS EN 1996-2*
BS 5642-1:1978+A1:2014 *Sills, Copings and Cappings. Specification for Window Sills of Precast Concrete, Cast Stone, Clayware, Slate and Natural Stone*
BS 5642-2:1983+A1:2014 *Sills, copings and cappings. Specification for copings and cappings of precast concrete, cast stone, clayware, slate and natural stone*
BS 6100-5.1:1992 *Glossary of Building and Civil Engineering Terms. Masonry. Terms Common to Masonry*
BS 6649:1985 *Specification for Clay and Calcium Silicate Modular Bricks*
BS 8000-3:2001 *Code of Practice for Masonry*
BS EN 1996-1-2:2005 *Eurocode 6. Design of Masonry Structures. General Rules. Structural Fire Design*

> BS EN 1996-2:2006 *Eurocode 6. Design of Masonry Structures. Design Considerations, Selection of Materials and Execution of Masonry*
> BS EN 1996-3:2006 *Eurocode 6. Design of Masonry Structures. Simplified Calculation Methods for Unreinforced Masonry Structures*
> BS EN 771-3:2011+A1:2015 *Specification for Masonry Units. Aggregate Concrete Masonry Units (Dense and Lightweight Aggregates)*
> BS EN 772-2:1998 *Methods of Test for Masonry Units. Determination of Percentage Area of Voids in Masonry Units (By Paper Indentation)*
> BS EN 934-3:2009+A1:2012 *Admixtures for Concrete, Mortar and Grout. Admixtures for Masonry Mortar. Definitions, Requirements, Conformity and Marking and Labelling*

7.5 STRUCTURAL/CARCASSING: METAL/TIMBERWORK

Contract requirements

7.5.1 Quality assurance

The specification should be verified for applicable QA schemes and the A/EA's instructions obtained on any duties arising. Reference is made to the use of the Timber Research and Development Association (TRADA) scheme under Trussed rafters (see 7.5.14) and to the project specification.

Part of the normal duties of the Clerk of Works is to verify for conformity with the specifications and drawings. In the case of structural steelwork and timber framing, guidance on inspection and approval of fabrication in workshops should be sought from the A/EA. The Clerk of Works will be required to inspect condition on delivery to site and monitor the site works. Some site decisions may be delegated while others are reserved for the A/EA or the project engineer. In this situation it is important that clear written instructions on the delegation of duties are obtained from the A/EA.

Quality control: structural steelwork

7.5.2 Fabrication

When inspection of fabrication in workshops, either on or off site, is carried out by others, the Clerk of Works may be required to provide liaison between them and the contractor for notification in good time of readiness of test pieces, forwarding drawings and other documentation as required by specification, and submitting test results for approval. A record of such actions should be made in the site diary.

Where responsibility for inspecting fabrication rests with the Clerk of Works, the delegation of duties should be established with the A/EA. Tolerances for accuracy of alignment, level and verticality are normally as given in BS EN 1090-2:2008+A1:2011, and for lengths, as specified. The setting out of holes and cuts should be verified.

7.5.3 Handling and transportation

To prevent primitive practices of handling and transportation it is now common to require a method statement not only for handling but also for the erection of components. The requirements are to be strictly enforced. Both the main contractor and the steelwork contractor should be informed in writing of any case of non-compliance, with copies to the A/EA and also to the health and safety coordinator if there are safety implications.

The Health and Safety Plan and the health and safety coordinator should be consulted in the case of any uncertainty over the methods being used.

Damage: the Clerk of Works must be strict in the rejection of components found to be damaged on delivery to site, rejecting without question dented hollow sections and distorted flanges of compression members, and insisting on removal of oil and grease.

Repairs: any proposals for site repair of damage should be reported to the project SE via the A/EA. If not rejected, details should be agreed with the A/EA as necessary.

7.5.4 Steel quality

The A/EA will have approved the method of steel manufacture, but the Clerk of Works may be required to inspect deliveries of steel for conformity of surface quality to BS 7668, BS EN 10029, BS EN 10025-1, BS EN 10025-2 or BS EN 10210-1 and BS EN 10210-2. Any steel with defects beyond the permissible limits in these standards, or where pitting by surface rust is excessive and visible to the naked eye, should be rejected.

7.5.5 Steel identification

Steel must be correctly identified with quality and other marks as specified, and these marks transferred to any cut sections. There should be an erection sequence drawing on site showing the location of members, and members should be marked in accordance with it.

7.5.6 Site fabrication

The section on Workmanship at 7.5.10 below provides additional information. The following are typical matters that may require the Clerk of Works' decision.

Plane table for fabrication: if used, it is important to ensure that it is level in both directions to guarantee accuracy of fabrication.

Temporary attachments: welding or bolting is not permitted in tension areas. If there is any doubt about whether an area is in tension, the A/EA or project SE should be consulted.

Cutting: any proposal for cutting other than by sawing, shearing or machine flame should be referred to the A/EA.

Faying surfaces: machining of deformed surfaces is not allowed without the A/EA's approval.

Holing: any members with incorrect holing are to be rejected.

Welding: this is normally confined to the workshop, but may be specified for some site works. Detailed instructions should be obtained from the A/EA. Certification of the welder's competence for the class of work as directed should be obtained before work starts.

Burrs: any burrs or sharp areas are to be removed by grinding, as these can reduce the durability of surface protection.

Indented anchor bolts: these are to be used only for locating, not for load.

High-strength friction grip (HSFG) bolts: these must be tightened only by use of indicating washers and not by torque wrench.

Holding down bolts: joint holes must be aligned prior to the installation of bolts. Bolts should never be driven into position, as this can damage threads or deform the bolt. Hot-dip galvanised steel is normally acceptable for Grade 8.8 bolts unless stainless steel is specified. Galvanising should be carried out by the bolt manufacturer.

Fit of nuts: the fit should be verified as specified. The aim is to avoid zinc-to-zinc contact between nut and bolt threads to reduce the danger of the nut seizing. Internal threads are protected by close contact with the galvanised threads of the bolt.

Stress relieving: submission of stress relief certification to the A/EA and galvaniser should be confirmed.

Galvanised steel: the Clerk of Works should check for transit damage and seek approval for remedial works. For cutting and welding of galvanised steel, the SE's approval and directions should be obtained and the Health and Safety Plan and the health and safety coordinator referred to in relation to associated health hazards.

7.5.7 Steel erection

Setting out: setting out must be verified.

Safety: the Clerk of Works should verify compliance with the Health and Safety Plan and the BCSA code of practice for erection of low-rise buildings, and ensure that the provision of access for inspection is satisfactory: a separate standard is applicable to that for erectors. The A/EA's directions should be obtained on allowing additional bolt holes or weldments to the permanent structure.

Stability: the contractor is responsible for ensuring the stability of the structure at all stages of construction and for the proper sequencing of stages of construction, taking into account other structural elements. The Health and Safety Plan should be examined for prescribed safe sequences of erection. The Clerk of Works should bring immediately to the attention of the contractor, A/EA and health and safety coordinator any departure from the method statement or any other potentially unsafe situations. Any further instructions can be obtained from the A/EA.

Cutting and drilling: site cutting and drilling is not normally permitted because it is seldom done satisfactorily and weakens surface coatings, but is sometimes unavoidable. Any proposal should be referred to the A/EA for authorisation and the A/EA's method statement, laying down conditions for such work obtained.

Note: at no time will the contractor be allowed to form holes with a burning torch.

Distortion: distortion of steelwork in the course of erection may exceed design stresses. Any instance should be reported to the A/EA.

Weather: if instability is suspected due to severe weather conditions, the A/EA and health and safety coordinator should be informed and the contractor instructed to verify the safety of the work.

Column bases: casting dates for maturity to accept erection loads should be confirmed. The A/EA's directions should be obtained on approval for raising or lowering column bases beyond specified limits, and for subsequent packing or grouting. The concrete must be sloped away from the embedded steel and the steel must not stand in water. The holding-down bolt pockets must be cleaned out and ready for grouting up before the steel columns are positioned over them.

Tolerances: the Clerk of Works should verify with the structural designer and the A/EA and confirm any particular tolerances specified. The fitting of gantry beams and crane rails should be verified against those required by the manufacturer of the traveller.

Temporary fixings: no temporary fixings that might damage protective surfaces are to be attached to the structure. The contractor's proposals should be presented to the A/EA for approval in advance.

Hoisting: the Clerk of Works should obtain and forward to the A/EA and health and safety coordinator details of any proposed method of lifting and winching using equipment fixed to structural members, identifying the members clearly and, if approved, monitoring any loading restriction. Care should be taken with loads from cranes. The steel erectors must use guidelines or ropes to control the swinging of loads during hoisting.

Phased connections: the A/EA's agreement and/or any restrictions for any proposed phasing must be obtained.

Holes: any misalignment should be reported to the A/EA. If not rejected, consideration should be given to the necessity for a larger bolt and the A/EA's agreement obtained.

Spanners: imperial spanners may not be used to tighten metric nuts and bolts.

HSFG bolts: the use of HSFG bolts requires a high level of supervision and inspection. The average gap of 0.25mm to which an HSFG bolt should be tightened relates to galvanised bolts. For other finishes the gap should be specified. BS EN 1993-1-8:2005 provides further information.

The correct washers must be used in assemblies. Over-tightening can damage bolts, and they could subsequently fail from brittle fracture in cold weather. When the gap of more than one fastener is zero and the average gap of the connection is as specified, the Clerk of Works should inform the A/EA, who should instruct replacement of the assembly (this may be a more rigorous requirement than the manufacturer's recommendations). Any fasteners that need to be slackened after full tightening should be discarded and not reused.

7.5.8 Inspection, certification and testing

Formal inspection, certification and testing are generally of a specialised nature, and the instructions of the A/EA should be obtained on any particular duties. These may include ensuring that samples, trial assemblies, test specimens, etc. are prepared and submitted in good time for the A/EA's attention, and forwarding the results of the tests to the A/EA. A record of all such actions should be kept. The Clerk of Works will normally be required to attend and supervise site tests, such as load testing in accordance with BS EN 1993-1-1:2005+A1:2014 Eurocode 3, unless the project engineer does so.

The Clerk of Works must ensure that the structure is complete before it is loaded for testing. This applies particularly to crane rails. Any problem should be reported to the A/EA and health and safety coordinator.

Inspection at workshops should always be carried out in the presence of the contractor, and the A/EA will normally require a report.

Particular attention should be paid to the production in good time of all material test certificates. The Clerk of Works should identify these from the specification, verify that they are submitted, forward them to the A/EA and keep a record.

Where there is any reason to doubt the quality of steel supplied, the A/EA may instruct the contractor to carry out special tests. Any suspicions should be reported immediately to the A/EA.

7.5.9 Protective coating

The Clerk of Works should verify that the contractor orders protective coating materials as specified. Paint containers should be checked immediately after delivery, and accepted only if they are labelled as described and within shelf life.

Where a poor quality of paint is suspected, the contractor will be asked to report this to the supplier and inform the A/EA, with details of supplier and material as on the container label. An unopened container should be set aside for testing.

Other materials: the Clerk of Works should agree delegation by the A/EA for approval of other materials, for example:

- rust removers
- detergent.

Decorative painting of zinc-rich coatings: the contractor must submit details in advance by the time specified for the A/EA's approval.

Cement wash: the A/EA is to instruct if rainwater from rusty steel is likely to damage finishes before embedding in concrete.

Sampling and testing: each batch of paints is to be sampled in accordance with BS 3900. The A/EA's instructions should be obtained on the number of samples required, and whether unopened containers are to be despatched to the nominated testing authority, or samples only from containers. Each different batch should be identified on delivery. Arrangements should be made for the contractor to set aside containers as appropriate and despatch for testing, with subsequent test results forwarded to the A/EA.

The methods to be followed in sampling from containers, paint kettles and finished work are detailed in BS EN ISO 15528:2013. Samples should be taken from paint kettles and finished work if there is reason to verify that the correct paint is being applied, and if faults occur during or after application.

Paint thickness: the A/EA's instructions should be obtained on the number and locations of measurements to be carried out using an elcometer (electromagnetic meter). The correct calibration of the meter must be confirmed before use. Coats of paint must be compatible with other coats, for example if intumescent paints are used then primer coats must be compatible.

Sample panels: the contractor must prepare sample panels of the specified finishes in good time for the A/EA's approval. Approved panels should be kept on site for comparison purposes.

7.5.10 Workmanship

Attention is drawn to the following points of workmanship.

Barrier coatings: aluminium coatings must never be allowed to have direct contact with rain-washed concrete, concrete in wash-down areas, brickwork, plaster or plaster products. A barrier coat of black bitumen is required. Dissimilar metals should also be separated by a barrier coat or sleeve of bitumen plastics. Steelwork embedded in brickwork should be protected with black bitumen. The A/EA should be consulted in any case not covered by specification.

Welding of galvanising: the limits in specification on sizes of contact areas relate to the risk of explosion in enclosed areas if welding is to be carried out. The Clerk of Works should verify the Health and Safety Plan for method statements, and inform the A/EA, health and safety coordinator and contractor immediately should any problems become evident.

Faying surfaces: if the contractor has a standard method of protecting faying surfaces equal to or better than the method specified, it may be accepted.

Repair of damaged zinc or aluminium coatings: if the method does not satisfactorily clean down to bare steel, grit blasting to SA2.5 (semi bright) may be authorised. Sealant needs to be removed from overlap areas to obtain metallic contact.

Colours: pigments are not to be added on site to tint zinc-rich coatings. Colours are to be prepared by the paint manufacturer.

Stripe coats: where zinc-rich paint is applied to welds and edges of steelwork in stripe coats, the Clerk of Works should verify that thorough preparation as specified is carried out, paying particular attention to the thickness of the coat at the edges as well as to overlaps as specified.

Pot life: the Clerk of Works should verify that mixed multi-component materials are used within pot life as recommended by the manufacturer.

Protection of fasteners: the A/EA's approval must be obtained for the contractor's proposed method.

Quality control: structural timber

7.5.11   Quality assurance

The Clerk of Works should verify the specification for any QA schemes applicable and obtain the A/EA's instructions on any duties arising. Reference is made to use of the TRADA scheme under Trussed rafters (see 7.5.14).

Quality control: it is part of the Clerk of Works' normal duty to verify for conformity with the specifications and drawings, to inspect condition on delivery to site and to supervise site work. Some site decisions may be delegated to the Clerk of Works while others are reserved for the A/EA or the project engineer. In this situation it is important that clear instructions on the delegation of duties are obtained from the A/EA.

7.5.12   Structural timber

All timber for structural purposes should be specified in the contract documentation to stress grades. The Clerk of Works must verify the project specification and that delivered timber is marked with its stress grading and used as specified. The A/EA should be informed of any unmarked or suspect materials. Timber should have had a preservative treatment applied – normally tanalised timber will have treatment documentation. TRADA publishes a guide to stress-graded softwood.

7.5.13   Timber frame buildings

It is vitally important to check that all horizontal cavity barriers are installed correctly to prevent the spread of fire through the cavity. The A/EA must be informed of any non-compliance.

7.5.14   Trussed rafters

TRADA: trussed rafters are subject to a QA scheme, and may only be supplied by a member of the Trussed Rafter Quality Assurance Scheme operated by TRADA Quality Assurance Services Limited. The supplier should be verified through the A/EA or by direct enquiry to TRADA (see Chapter 17 for contact information). Marking will be either:

● a stamp or label stating that the supplier is a member of the Trussed Rafter Quality Assurance Scheme, or
● a stamp or label with the trademark and the scheme mark.

Unloading, handling and workmanship generally should be specified to comply with recommendations in the Trussed Rafter Association's *(TRA) technical handbook*, and guidance on erection is contained in BS EN 1995-1-1:2004+A2:2014 Eurocode 5. The limitations to the TRADA QA scheme should be made clear, in that the building designer is held responsible for the overall design and structural integrity of the building, a roof designer may undertake only detailed design of the roof structure, and the trussed rafter designer is responsible only for the design of the trussed rafter.

Deliveries should be checked for any damage, and damaged rafters returned to the supplier; rafters should not be repaired on site unless the A/EA has given permission after the manufacturer has specified the repair and issued a method statement for site work.

The following points should be verified, particularly during storage and erection:

● stress grading marks
● storage clear of the ground, and protection from weather
● verticality ensured during storage, lifting and installation
● rafters to be lifted only at the manufacturer's lifting points, which should be marked on the rafters

- sufficient bracing provided during erection, and permanent stability and wind bracing on completion
- satisfactory fixing to the building structure
- no cutting of any member (excluding overhangs)
- no site repairs in the event of damage, except as approved by the manufacturer (see above)
- if gang-nail plates are to be used, they must have 'nailable'-type nail holes for the sections to be joined, and correct length nails are to be used.

Shop drawings: where required, these must be obtained from the contractor at least four weeks before the programmed fabrication date and submitted to the A/EA for approval.

Testing: the A/EA's instructions for testing should be obtained. Normally a prototype rafter is load tested, and the A/EA may require testing of water content of timber.

Statutory requirements and technical standards

Publications to which reference may be necessary include:

Steelwork

BS 3900-0:2010 *Methods of Test for Paints. Index of Test Methods*

BS 5493:1977 *Code of Practice for Protective Coating of Iron and Steel Structures Against Corrosion* [Partially replaced by BS EN ISO 12944 Parts 1 to 8 1998 Paints and Varnishes, and BS EN ISO 14713:1999 Protection Against Corrosion of Iron and Steel in Structures. Zinc and Aluminium Coatings. Guidelines]

BS 7668:2016 *Weldable Structural Steels. Hot Finished Structural Hollow Sections in Weather Resistant Steels. Specification*

BS EN 10025-1:2004 *Hot Rolled Products of Non-alloy Structural Steels. General Delivery Conditions*

BS EN 10025-3:2004 *Hot Rolled Products of Non-alloy Structural Steels. Technical Delivery Conditions for Long Products*

BS EN 10025-4:2004 *Hot-Rolled Products of Structural Steels. Technical Delivery Conditions for the Thermomechanical Rolled Weldable Fine Grain Steels*

BS EN 10029:2010 *Hot-Rolled Steel Plates 3mm Thick or Above. Tolerances on Dimensions and Shape*

BS EN 10210-1:2006 *Hot Finished Structural Hollow Sections of Non-alloy and Fine Grain Steels. Technical Delivery Requirements*

BS EN 1993-1-1:2005+A1:2014 *Eurocode 3. Design of Steel Structures. General Rules and Rules for Buildings*

BS EN 1993-1-10:2005 *Eurocode 3. Design of Steel Structures. Material Toughness and Through-Thickness Properties*

BS EN 1993-1-5:2006 *Eurocode 3. Design of Steel Structures. Plated Structural Elements*

BS EN 1993-1-8:2005 *Eurocode 3. Design of steel structures. Design of joints* (replaced both BS 4604 Parts 1 and 2)
BS EN 1993-5:2007 *Eurocode 3. Design of steel structures. Piling*
BS EN 1993-6:2007 *Eurocode 3. Design of steel structures. Crane supporting structures*
BS EN ISO 15528:2013 *Paints, varnishes and raw materials for paints and varnishes. Sampling*
British Constructional Steelwork Association (BCSA), *BCSA code of practice for erection of low rise buildings, 2004*

Timber

BS 8000-5:1990 *Workmanship on building sites. Code of practice for carpentry, joinery and general fixings. Partially replaced by BS 8000-0:2014 Workmanship on construction sites. Introduction and general principles*
BS EN 1995-1-1:2004+A2:2014 *Eurocode 5: Design of timber structures. General. Common rules and rules for buildings*
BS EN 1995-1-1:2004+A2:2014 *Eurocode 5: Design of timber structures. General. Common rules and rules for buildings*
BS EN 594:2011 *Timber structures. Test methods. Racking strength and stiffness of timber frame wall panels*
BS EN ISO 2884-1:2006, BS 3900-A7-1:2006 *Paints and varnishes. Determination of viscosity using rotary viscometers. Cone-and-plate viscometer operated at a high rate of shear*
BS EN ISO 2884-2:2006, BS 3900-A7-2:2003 *Paints and varnishes. Determination of viscosity using rotary viscometers. Disc or ball viscometer operated at a specified speed*
British Constructional Steelwork Association (BCSA), *BCSA Code of Practice for Erection of Low Rise Buildings, 2004*
TRADA *Timber Strength Grading and Strength Classes, 2003. Trussed Rafter Association (TRA) TRA technical handbook*

7.6 CLADDING/COVERING

Contract requirements

7.6.1 Duties

It is part of the Clerk of Works' normal duties to verify for conformity with the specifications and drawings. Cladding and coverings are often of a proprietary type, and it is important to verify their particular fixing requirements and tolerances. This aspect is considered further below.

7.6.2 Manufacturer's instructions

Many clauses in the specification refer to the need to store, handle and fix in accordance with the manufacturer's instructions. Such references should be identified, copies of these instructions obtained and compliance confirmed, including the use of sealants and other accessories as recommended by the manufacturer. Reference may also be made to compliance with application requirements of a relevant British Standard. If any doubts arise, the A/EA should be consulted.

7.6.3 A/EA's approval

The A/EA's instructions should be obtained on any approvals required in the specification that may be delegated in part or whole to the Clerk of Works. The following are examples:

- **patent glazing:** approval to cut or drill parts of the structure for fixing
- **metal profiled flat sheet:** evidence of compliance with fire rating classification
- **glass reinforced plastics:** approval to repair damage and if given, approval of method
- **precast concrete slabs:** approval of purpose-made lifting devices and testing of fixings for lifting; the A/EA may also specify test loading of a sample of panels; fixing may not normally be carried out until panels reach their 28-day strength, unless the A/EA agrees otherwise
- **natural stone slab:** supplier's guarantee that stone is from the quarry and bed specified
- **natural slating:** supplier's certificate of compliance with atmospheric pollution resistance requirements; the Clerk of Works should verify whether slates are to be fixed via nailing or clipping
- **lead sheet:** laying other than in accordance with BS 6915
- **aluminium sheet:** laying other than in accordance with CP 143-15
- **copper sheet:** laying other than in accordance with CP 143-12.

Quality control: slating and tiling of roofs

7.6.4 Quality assurance

The specification should be verified for applicable QA schemes and the A/EA's instructions on any duties arising obtained.

7.6.5 Warning notices

Warning notices (e.g. 'Fragile roof covering – use crawling boards') are to be supplied by the contractor. Orders are to be placed in time before completion and must conform to HSE standards. Notices will be required as part of the contractor's own safe working precautions.

7.6.6 Materials

Delivery documents should be verified for evidence of compliance with particular performance requirements such as thermal conductivity, fire performance, light diffusion class, etc. The contractor should be asked to provide additional certification as necessary. In case of any doubt, the A/EA should be consulted.

7.6.7 Samples

The specification should be verified for requirements to submit samples, and these should be submitted in good time for the A/EA's approval. Approved samples are to be kept on site to verify against deliveries.

7.6.8 Fabrication drawings

The specification should be verified for requirements to submit fabrication drawings, and these are to be submitted by the date specified for the A/EA's approval. A copy of the approved drawing(s) should be kept and the A/EA notified of any variation or discrepancy occurring in construction.

7.6.9 Approved firms

Where the specification provides for supply and/or fixing by approved firms, the Clerk of Works should verify that these firms do in fact carry out the service. It should be noted that precast concrete cladding units are to be obtained from a firm on the BSI register of firms of assessed capability.

7.6.10 Off-site inspection

The Clerk of Works will provide liaison, if required, for off-site inspections.

7.6.11 Safety precautions

The contractor is responsible for preventing unsafe access and egress to areas where roofing work is in hand and for ensuring the use of crawling boards as necessary. The contractor's attention should be drawn to any unsafe/dangerous practice, and the health and safety coordinator and A/EA informed and a record made in the site diary.

7.6.12 Workmanship

The specification should be consulted as to the type of underfelt to be used and the manufacturer's instructions for the required laps. Battens should have preservative treatment applied and must be laid with no more than three straight joints to a single rafter, with no battens less than 1200mm in length laid. The specification for slates or tiles should be verified – whether they are to be clipped or nailed and, in the case of tiles, how many and which courses should be fixed. Generally, the eaves, ridge line and edges to gables, hips and valleys are fixed, as well as every fourth or fifth course, depending on the roof pitch. If any doubt arises, the matter should be referred to the A/EA.

Statutory requirements and technical standards

Publications to which reference may be necessary include:

BS 8000-6:2013 *Workmanship on Building Sites. Code of Practice for Slating and Tiling of Roofs and Claddings*
BS 5534:2014+A1:2015 *Slating and Tiling for Pitched Roofs and Vertical Cladding. Code of Practice*
BS 6100-6:2008 *Building and Civil Engineering. Vocabulary. Construction Parts*
BS 6915:2001+A1:2014 *Design and Construction of Fully Supported Lead Sheet Roof and Wall Coverings. Code of Practice*
CP 143-12:1970 *Code of Practice for Sheet Roof and Wall Coverings. Copper. Metric Units*
CP 143-15:1973 *Code of Practice for Sheet Roof and Wall Coverings. Aluminium. Metric Units*

7.7 WATERPROOFING

Contract requirements

7.7.1 Duties

It is the normal duty of the Clerk of Works to verify conformity with the specifications and drawings. Specialist subcontractors are involved, and work prepared by one contractor to be finished by another must be carefully verified. It is becoming more common to obtain confirmation in writing from the asphalter before asphalting work starts that the preparation of areas to be covered is acceptable. This avoids any opportunity of claiming that defects in roofing were due to poor preparation. In certain areas a 'permit to work' may be required for operatives.

Quality control

7.7.2 Quality assurance

The specification should be verified for applicable QA schemes and the A/EA's instructions obtained on any duties arising. Use is made in this workgroup of BSI Kitemarked products.

7.7.3 Materials

Delivery documents should be verified for evidence of compliance with particular performance requirements, such as thermal conductivity. The contractor should be asked to provide additional certification as necessary. Unmarked materials should be rejected. The A/EA should be consulted if any doubt arises. Kitemarks should be verified where specified. In cases where the contractor offers materials as equivalent to those specified, these should be referred to the A/EA for approval or otherwise.

7.7.4 Roofing

Bonding: where a high bond primer is required for cementitious or masonry surfaces, the Clerk of Works should verify that the primer is acceptable to the asphalter.

Mastic asphalt or felt roofing: written agreement by the roofing contractor that the base and falls are satisfactory for work to start must be obtained.

Cauldrons: the A/EA's instructions on the acceptability of using cauldrons should be obtained.

Thermometer: the contractor should have an appropriate thermometer on site and should use it to observe specified temperature limits when laying asphalt.

7.7.5 Asphalt samples

The specification should be verified for requirements to test samples of asphalt. These must be taken in accordance with BS 5284 section 4 as specified, and despatched to an independent laboratory for analysis. Analyses received should be forwarded to the A/EA. The contractor is required to keep a record of locations; the Clerk of Works should verify that this is done, but make a separate record in the site diary. It should be noted that BS 5284 requires sampling to be carried out at the time of laying.

7.7.6 Liquid applications

The Clerk of Works should verify for any temperature restrictions in the manufacturer's instructions where damp-proof membranes are applied in liquid form, and monitor compliance.

7.7.7 Protection against traffic

The contractor is responsible for ensuring that traffic on felted roofs does not exceed the design category. The A/EA should be asked to confirm what this load limit is, and compliance should be monitored. It is also important that the contractor keeps all traffic on the finished roofing to a minimum to prevent damage.

Statutory requirements and technical standards

Publications to which reference may be necessary include:

BS 747:2000 *Reinforced Bitumen Sheets for Roofing. Specification*
BS 1446:1973 *Specification for Mastic Asphalt (Natural Rock Asphalt Fine Aggregate)
for Roads and Paths*
BS 1447:1988 *Specification for Mastic Asphalt (Limestone Fine Aggregate) for Roads
and Footways*
BS 5284:1993 *Methods of Sampling and Testing Mastic Asphalt Used in Building
and Civil Engineering*
BS 8000-4:1989 *Workmanship on Building Sites. Code of Practice for Waterproofing
Mastic Asphalt Council (MAC) Roofing Handbook*

7.8 LININGS/SHEATHING/DRY PARTITIONING

Contract requirements

7.8.1 Duties

It is part of the normal duties of the Clerk of Works to verify for conformity with the specifications and drawings. Where panels of similar appearance but of different performance requirements are being installed, it is important to verify that the correct types are used.

Quality control

7.8.2 Quality assurance

The specifications should be verified for applicable QA schemes, and the A/EA's instructions obtained on any duties arising. Schemes that may be used in this workgroup include Agrément certification.

7.8.3 Material

The delivery documents should be verified for certification of edge and other treatments and fire performance, as specified. The contractor should be asked to provide additional certification as necessary. The A/EA should be consulted if any doubt arises.

7.8.4 Manufacturer's instructions

Requirements in the specification to fix in accordance with the manufacturer's instructions or recommendations should be identified, copies of these obtained and compliance monitored.

7.8.5 Samples

The specification should be verified for requirements to submit samples, and these should be submitted in good time for the A/EA's approval. Approved samples should be kept on site for verification against deliveries.

7.8.6 Workmanship

BS 8000-8 and BS 8000-0 apply. It should be remembered that double-boarded partition walling should have staggered joints, for example.

Plasterboard dry lining: metal angle bead reinforced angles should be used where external angles need maximum protection. Taper-edged boards are to be used where flush finish joints are required.

Veneers: Refer to BS EN 14374:2004 and BS EN 14279:2004+A1:2009 for definitions, classifications and specifications. The A/EA's instructions on direction should be obtained.

7.8.7 Loose insulants

Loose insulants should be used and placed correctly to achieve the specified requirement for sound or thermal insulation.

7.8.8 Demountable partitions

The setting out of partition centre lines by the subcontractor and the marking out of positions of vertical abutment to structure by the main contractor should be verified.

7.8.9 Raised access floors

The specification may require supply and fixing by an Agrément-certificated contractor, be it computer flooring, timber stage flooring or access duct panels. The information required from the manufacturer will normally have been submitted at tender stage.

7.8.10 Tests

Site testing is required for earth bonding continuity, in liaison with the M&E site inspector. Results should be forwarded to the A/EA. The Clerk of Works should carry out any other tests of demountability and transferability as required by the A/EA.

Statutory requirements and technical standards

Publications to which reference may be necessary include:

BS 8000-8:1994 *Workmanship on Building Sites. Code of Practice for Plasterboard Partitions and Dry Linings*
British Gypsum, *The White Book*
Lafarge Roofing, *The Red Book*

7.9 SURFACE FINISHES

Contract requirements

7.9.1 Duties

It is part of the normal duty of the Clerk of Works to verify for conformity with the specifications and drawings.

7.9.2 Quality assurance

The specifications should be verified for applicable QA schemes and the A/EA's instructions obtained on any duties arising.

7.9.3 Materials

Delivery documents should be verified for evidence of compliance with particular performance requirements or limitations (e.g. admixtures). The contractor should be asked to provide additional certification as necessary. The A/EA should be consulted if any doubt arises.

Compliance with the specifications and drawings should be monitored. The A/EA should be informed and the A/EA's approval obtained for any alternative products proposed by the contractor before their use is allowed. Any delegation by the A/EA to approve materials, such as proprietary paints and coatings and materials offered by the contractor as equivalent to those specified, should be approved.

7.9.4 Manufacturer's instructions

The specification should be verified for reference to installation in accordance with the manufacturer's instructions, copies of these obtained and compliance checked.

7.9.5 Specialist firms

Where the contractor is required to select from listed firms for specialist work, the Clerk of Works should verify that the work is in fact carried out by an approved firm.

7.9.6 Samples

The specification should be verified against requirements to submit samples or to prepare sample panels, and the Clerk of Works must ensure that these are submitted in good time for the A/EA's approval. Approved samples of materials should be kept on site for comparison.

Quality control: cement-sand floor screeds and concrete floor toppings

7.9.7 Quality assurance

The specification should be verified against product QA if applicable. For screeding, cement is to be supplied by a BSI-registered firm.

7.9.8 Materials

Delivery notes should be verified for evidence of compliance with performance requirements or limitations (e.g. admixtures). The contractor may be asked to provide additional documentation if necessary. Any offer of equivalent materials to those specified should be referred to the A/EA for approval.

7.9.9 Items for specific attention

Aerated cement screeds: for proprietary lightweight aerated cement screeds where the structural slab is laid to falls of 1:40, the screeds should be laid to a consistent thickness not exceeding 50mm. Where the structural slab is level, the screed should be laid to falls with a maximum thickness of 200mm. Where this means that a fall of 1:40 is not obtainable, a lesser fall is acceptable — except for asphalt covering, which should not be less than 1:80. Any difficulty should be reported to the A/EA.

Water: use of any source other than mains supply is to be approved. A sample analysis should be obtained in case of any doubt.

Concrete bases: preparation of hardened smooth concrete by mechanical scabbling may damage bases of less than 100mm thickness. In such cases preparation should be by shot or grit blasting. If the contractor proposes water pressure hoses, the A/EA should be consulted.

Monolithic screeds: these are specified to be laid within three hours of laying base concrete, but in hot weather a lesser period may be agreed with the contractor. In the absence of agreement the A/EA should be consulted.

Curing: the A/EA is to approve any alternative method of curing proposed by the contractor.

Impact test — dense floor screeds: the A/EA should determine locations for test and any delegation of approval if the British Standards centres and specific testing parameters are to be varied. The Clerk of Works should witness tests and record readings on drawings and in the site diary. For acceptance, other defects are to be taken into account such as curling, excessive cracking of material and breakdown of hollowness. Additional tests may be ordered as necessary. The A/EA's instruction or agreement for remedial action should be obtained where required.

Screed tolerances: project specification should be verified for finished tolerances; normally, 3mm in 3m.

Quality control: plaster and cement rendering

7.9.10   Quality assurance

The specification should be verified to see if Product Quality Assured items have been requested. All items covered by this scheme will be accompanied by a certificate.

7.9.11   Quality control

Conformity with the specifications and drawings should be verified. No additives are to be included in mortar mixes without prior approval from the A/EA.

7.9.12   Special points to watch

Suction: any excessive suction must be addressed either by means of damping down with clean water or by sealing the surface with a bonding agent prior to the first coat of render being applied.

Shrinkage: some forms of lightweight block when saturated shrink to a great extent on drying and cause render to debond. It is the Clerk of Works' responsibility to ensure that this form of blockwork does not get excessively wet during storage or construction.

Cement: supplies must come from a BSI-registered firm.

Water: use of a source other than mains supply is to be approved. In cases of doubt a sample should be sent for analysis.

Sample areas: preparation of sample areas of external render finishes must be undertaken in good time for the A/EA's approval before work proceeds.

Surface salts: where surface salts repeatedly appear on masonry or concrete backgrounds despite dry brushing as specified, the A/EA's instruction should be obtained.

Mechanical application: the A/EA's approval is required if the contractor proposes to use machine-applied plaster. If approved by the A/EA, the plaster must be applied by means acceptable to the plaster manufacturer and by experienced operatives.

Internal working areas: the Clerk of Works must ensure that there is enough natural or artificial light for tradespeople to carry out their work correctly. Any areas with inadequate lighting should be reported to the A/EA.

Surface finish: an accuracy of finished surface within 3mm is required under a 1.8m straight-edge for smooth untextured plaster unless otherwise specified; refer to the project specification. The A/EA's instruction must be obtained for any plaster finish other than trowelled smooth.

Formed joints: the A/EA's approval is required for any cutting of formed joints instead of using scrim reinforcement at junctions of gypsum baseboard and solid background.

Scraped finish: where the surface is to be scraped to an approved finish the A/EA's instructions must be obtained. Scraping removes about 3mm of coating.

Quality control: wall and floor tiling

7.9.13 Quality control

The Clerk of Works should verify and monitor conformity with the specifications and drawings, and verify that products are installed in accordance with the manufacturer's instructions.

7.9.14 Special points to watch

Samples: the A/EA's approval must be obtained prior to the ordering. Approved samples should be kept on site for comparison.

Sample areas: preparation of sample areas in locations as directed by the A/EA must be undertaken in good time for the A/EA's approval before work begins.

Tolerances: if, on new concrete, specified tiling tolerances cannot be met, then the A/EA's approval must be sought for rendering the concrete background.

Impact protection: certain wall tile areas, such as corners, pillars and those near to door openings, may be subjected to impact. These locations can be protected by securing metal angles over the tiling or building in special recessed corner angles. Any gaps or crevices caused by securing these metal angles should be completely filled, especially if in a hygienic area.

Falls: adequate falls should be incorporated in floor areas likely to come into contact with acidic residues, such as lactic acid in dairies, and detergent spillages. Gradients between 1:80 and 1:40 are recommended. The direction of falls should be planned with the traffic flow in mind, so that the traffic will move across rather than up and down the slope. The position of drainage channels and gulleys should be given special attention.

Substrates: unusual substrates (in existing buildings) may be encountered and specialist advice may be required from the A/EA and specialist adhesive manufacturers.

General design considerations are discussed in section 3 of BS 5385: Parts 1 to 5.

7.9.15 Movement joints

Movement joints should be provided in accordance with BS 5385-1 for walls and in accordance with BS 5385-3 for floors. Their type and location should have been decided at the design stage. Each movement joint should be at least 6mm wide and of a depth at least equal to the thickness of the tile and bedding. Movement joints should be impervious and the sealant well bonded to the sides of the joints.

It is important that any movement joints already incorporated in the structure should not be tiled over but be carried through to the face of the tiling.

In large areas of wall tiling, movement joints should be provided at internal vertical corners and at 3m to 4.5m centres horizontally and vertically.

Movement joints should be provided around the perimeter of the floor and where floor tiles abut fixed machinery and structural fixtures such as columns, bases, etc. Intermediate movement joints should be incorporated in large floor areas, as described in BS 5385-3.

Brass or stainless steel reinforced movement joints using bonded neoprene should be used for intermediate joints, especially where there is the likelihood of wheeled traffic.

All other movement joints should be filled completely with appropriate sealant.

7.9.16   Materials

Cement: cement should comply with the requirements of BS EN 197-1:2011, or BS EN 413-1:2011.

Sand: sand for rendering should comply with BS EN 13139:2002. Sands for cement/ sand screeds and mortar beds should comply with grading in BS EN 12620:2013.

Water: use of source water other than mains supply is to be approved. In case of any doubt a sample should be sent for analysis. Water should be fresh and clean.

Tiles: project specification should be verified, but wall tiles are to comply with BS EN 14411:2016. Floor tiles are to comply with BS EN 14411:2016 and should be selected to suit service conditions.

Adhesives and grout: the most frequent failures occur because of the use of the wrong combination of adhesive and grout for the service conditions. The Clerk of Works should check particularly for adequate water resistance of both, for example adhesives to comply with the performance requirements of BS EN 12004-1:2017 possessing Class AA water resistance.

Quality control: decorative wall coverings and painting

7.9.17   Safety

The Health and Safety Plan should be examined for procedures and method statements relating to paint storage, handling and application. Any failure to observe safe methods should be brought to the contractor's attention and the health and safety coordinator should be informed. Compliance with mandatory safety clauses relating to the Control of Lead at Work Regulations 2002 and the need to ensure adequate ventilation for all painting processes or provide pressure-demand respirators should be monitored.

Danger areas

For areas where explosives or highly flammable materials are stored or handled, the specification should cover any restriction on materials, but it should be noted that aluminium paints are not to be applied to rusty surfaces in such situations. In case of doubt the A/EA should be consulted.

7.9.18 Sample areas

Sample areas of each type of coating should be prepared in good time for approval by the A/EA (or the Clerk of Works, if delegated) before the work proceeds. An area of about 1m^2 is sufficient, preferably in a corner of a wall where varying light and shade will give greater definition of colour than a central panel. Sample areas for joinery are best applied to a panel of plywood of similar finish and appearance to the article to be coated rather than to the article itself, to avoid the labour of cleaning it off.

7.9.19 Special points to watch

Background preparation: all paintwork will fail, regardless of quality of product, if the background is unsuitable, badly prepared or is contaminated with oil, grease or even excessive amounts of moisture. Correct background preparation should be monitored. Oil or grease spots may be touched over with a 'styptic' paint, such as aluminium paint, before undercoat and topcoat are applied. This should seal them in.

Labelling: paint containers should be checked immediately after delivery and accepted only if labelled as described in the specification. The requirement that all coats of each paint system are to be obtained from the same supplier should be noted.

Paint quality: where poor quality paint is suspected, the contractor should be asked to report this to the supplier and the A/EA should be informed with details of supplier and material as shown on the container label. An unopened container should be set aside for testing.

Flame-retardant paint: the Clerk of Works must ensure that certification as specified is obtained and forwarded to the A/EA. Primer coats must be checked for compatibility. (See also 7.5.9, above.)

Sampling and testing: the A/EA's direction must be obtained on sampling and testing. See also 7.9.18, above.)

Paint thickness: a wet paint thickness gauge (if available) should be used to verify applied thicknesses.

Batch matching: the batch numbers of wallpaper to be used on adjoining walls within a room should be verified. If differing shades do occur, this should be brought immediately to the attention of the contractor, the A/EA informed of any instances and a record made in the site diary.

Disposal of paint containers: many paints (especially lead and solvent-based) should be treated as 'special waste' and disposed of in segregated skips.

7.9.20 Workmanship

The following are matters that may require the A/EA's approval or decision, for which any delegations should be agreed.

Concrete surfaces: the A/EA's instructions should be obtained for removal of releasing agents if these cannot be removed by washing down.

Defective zinc coating: the A/EA's instructions should be obtained before painting. The preferred action is that the defective coating be made good by the subcontractor or supplier concerned. An inferior remedy is to wire brush the defective areas back to bare steel and prime immediately with primer as instructed by the A/EA.

Grease and dirt: the specification provides for washing off with detergent and warm water, but this should be done sparingly and only as necessary. Wet preparation should generally be avoided, particularly on plaster.

Unsound timber: the contractor is to inform the A/EA on discovery of unsound timber and await the A/EA's instructions.

Thinning: thinning of paints and varnishes is only permitted if recommended by the manufacturer or by agreement of the A/EA. If allowed, in the case of oil paint a limit of 5 per cent by volume of thinners is specified, but in the case of emulsion paint, the specified limit of 10 per cent by volume of water may be exceeded with the A/EA's approval.

Sealing: the A/EA's instructions are to be given where sealing as specified proves inadequate or may be harmful to subsequent paint coats.

Quality control: mastic asphalt

7.9.21 Samples for analysis

If specified, samples should be taken in accordance with BS 5284 section 4 and sent for independent analysis. All analyses received should be forwarded to the A/EA. The contractor is required to keep a record with locations and to verify that this is done, but the Clerk of Works should make a separate record in the site diary. It should be noted that BS 5284 requires sampling to be carried out at the time of laying.

7.9.22 Sample area

The A/EA's approval should be obtained before work proceeds, unless delegation to the Clerk of Works has been agreed.

7.9.23 Cauldrons

The A/EA's instructions on acceptability of using cauldrons must be obtained.

Quality control: terrazzo finishes

7.9.24 Special points to watch

BS 8000-11 gives detailed information.

Cement: must be supplied by a BSI-registered firm.

Water: use of a source other than mains supply is to be approved. A sample should be sent for analysis in case of any doubt.

Samples: samples are to be obtained from the specialist subcontractor for the A/EA's approval. Approved samples should be kept on site for comparison.

Sample areas: preparation of sample areas in locations as directed by the A/EA must be carried out in good time for the A/EA's approval before work begins.

Grinding: the A/EA's instructions on the use of any alternative method to the wet process must be obtained.

Quality control: sheet floor finishes – rubber, plastics, cork, linoleum and carpeting

7.9.25 Conditions prior to laying

Orders must be placed in adequate time for fixing.

The A/EA is to be satisfied that the building is weathertight, all wet trades have been completed, the building has dried out, paintwork is finished and dry and all fixtures abutting the floor covering have been fixed.

The contractor is to confirm that the flooring contractor's written agreement has been obtained that the base is suitable to receive the specified covering.

Tests for moisture content of screeds or substrate are to be taken prior to laying. The Clerk of Works should confirm acceptable relative humidity of storage and the laying area and obtain the assistance of the M&E site inspector to carry out tests as required.

7.9.26   Points to watch

Storage: many sheet finishes should be stored for a period of time prior to laying in the same conditions as the area to be covered. The supplier's and manufacturer's instructions should be consulted.

Sample areas: where specified, the A/EA should be informed when the sample area is ready for inspection.

Repairs: only minor repairs may be carried out by the contractor. Any case of serious damage should be referred to the A/EA for instructions.

Seams: seams are to be avoided on traffic routes. If unavoidable, they should run parallel to the route.

Statutory requirements and technical standards

Publications to which reference may be necessary include:

Screeds and concrete finishes

BS 6100-6:2008 *Building and Civil Engineering. Vocabulary. Construction Parts*
BS 8000-0:2014 *Workmanship on Construction Sites. Introduction and General Principles*
BS 8204-1:2003+A1:2009 *Screeds, Bases and in Situ Floorings. Concrete Bases and Cementitious Levelling Screeds to Receive Floorings. Code of Practice*

Rendering and plastering

BS 5385-1:2009 *Wall and Floor Tiling. Design and Installation of Ceramic, Natural Stone and Mosaic Wall Tiling in Normal Internal Conditions. Code of Practice*
BS 5385-2:2015 *Wall and Floor Tiling. Design and Installation of External Ceramic, Natural Stone and Mosaic Wall Tiling in Normal Conditions. Code of Practice*
BS 5385-3:2014 *Wall and Floor Tiling. Design and Installation of Internal and External Ceramic and Mosaic Floor Tiling in Normal Conditions. Code of Practice*
BS 5385-4:2015 *Wall and Floor Tiling. Design and Installation of Ceramic and Mosaic Tiling in Specific Conditions. Code of Practice*
BS 5385-5:2009 *Wall and Floor Tiling. Design and Installation of Terrazzo, Natural Stone and Agglomerated Stone Tile and Slab Flooring. Code of Practice*
BS 5493:1977 *Code of Practice for the Protective Coating of Iron and Steel Structures Against Corrosion. [Partially replaced by BS EN ISO 12944 Parts 1 to 8 1998 Paints and Varnishes and BS EN ISO 14713:1999 Protection Against Corrosion of Iron and Steel in Structures. Zinc and Aluminium Coatings. Guidelines]*
BS 6150:2006+A1:2014 *Painting of Buildings. Code of Practice*
BS 8000-0:2014 *Workmanship on Construction Sites. Introduction and General Principles*
BS 8000-0:2014 *Workmanship on Construction Sites. Introduction and General Principles*

BS 8000-11:2011 *Workmanship on Building Sites. Internal and External Wall and Floor Tiling. Ceramic and Agglomerated Stone Tiles, Natural Stone and Terrazzo Tiles and Slabs, and Mosaics. Code of Practice*

BS 8481:2006 *Design, Preparation and Application of Internal Gypsum, Cement, Cement and Lime Plastering Systems. Specification*

BS EN 12004-1:2017 *Adhesives for Ceramic Tiles. Requirements, Assessment and Verification of Constancy of Performance, Classification and Marking*

BS EN 12004-2:2017 *Adhesives for Ceramic Tiles. Test Methods*

BS EN 13914-1:2005 *Design, Preparation and Application of External Rendering and Internal Plastering. External Rendering*

BS EN 13914-2:2016 *Design, Preparation and Application of External Rendering and Internal Plastering. Internal Plastering*

BS EN 14411:2016 *Ceramic Tiles. Definition, Classification, Characteristics, Assessment and Verification of Constancy of Performance and Marking*

BS EN ISO 4618:2014 *Paints and Varnishes. Terms and Definitions*

PD CEN/TR 15123:2005 *Design, Preparation and Application of Internal Polymer Plastering Systems*

Mastic asphalt

BS 5284:1993 *Methods of Sampling and Testing Mastic Asphalt Used in Building and Civil Engineering*

Floor coverings

BS 5325:2001 *Installation of Textile Floor Coverings. Code of Practice*

BS 5385-3:2014 *Wall and Floor Tiling. Design and Installation of Internal and External Ceramic and Mosaic Floor Tiling in Normal Conditions. Code of Practice*

BS 5385-5:2009 *Wall and Floor Tiling. Design and Installation of Terrazzo, Natural Stone and Agglomerated Stone Tile and Slab Flooring. Code of Practice*

BS 8203:2017 *Code of Practice for the Installation of Resilient Floor Coverings*

Materials

BS EN 197-1:2011 *Cement. Composition, Specifications and Conformity Criteria for Common Cements*

BS EN 413-1:2011 *Masonry Cement. Composition, Specifications and Conformity Criteria*

BS EN 12004-1:2017 *Adhesives for Ceramic tiles. Requirements, Assessment and Verification of Constancy of Performance, Classification and Marking*

BS EN 12004-2:2017 *Adhesives for Ceramic Tiles. Test Methods*

BS EN 12620:2013 *Aggregates for Concrete*

BS EN 13139:2013 *Aggregates for Mortar*

BS EN 14411:2016 *Ceramic tiles. Definition, Classification, Characteristics, Assessment and Verification of Constancy of Performance and Marking*

7.10 FURNITURE, EQUIPMENT AND INTERNAL PLANTING

Contract requirements

7.10.1 Duties

It is a normal duty of the Clerk of Works to verify for conformity with the specifications and drawings. In the case of items obtained through the buying agency or client's direct supplier, their inspectors may carry out checks. For internal planting, the A/EA may arrange specialist inspections.

Quality control

7.10.2 Quality assurance

The specification should be verified for applicable QA schemes and the A/EA's instructions obtained on any duties arising. For supply of sanitary appliances and fittings, use is made of BSI Kitemark schemes.

7.10.3 Delivery of items for inclusion in the works

Together with the contractor, the Clerk of Works should inspect deliveries of items for compliance with the specification. Any damaged or missing items, which the contractor is required to record on the carrier's consignment note, should be agreed. The contractor must order replacements as necessary. A record should be kept in the site diary.

Specified requirements for maintenance of temperature and humidity conditions in storage and after installation should be noted. The Clerk of Works should verify these (with the assistance of the M&E site inspector) and report any problems arising to the A/EA.

7.10.4 Sanitary appliances: A/EA's approval

The following matters may require contractual approval.

Location of taps and valves: approved practice is to set the hot tap to the left to aid disabled and partially sighted people, and valves should be located in an accessible position.

Chemical closets: to be fixed in accordance with the manufacturer's instructions and as directed by the A/EA. The A/EA's instructions should be obtained.

Testing: performance testing of sanitary appliances is normally carried out when testing the drainage installation. Guidance on testing is given in BS 8000-0.

Compliance with water regulations: it is usually necessary to inspect for compliance with water regulations, particularly for correct provision of air-gaps and warning pipes. BS 8000-15 provides detailed guidance.

7.10.5 Internal planting

Where there is significant provision of internal planting, it is probable that the A/EA will arrange for specialist inspection of the work, particularly off site. The following are points that may require the attention of the Clerk of Works.

Certification: deliveries of plants and materials should be verified for labelling and certification to specified requirements. Verification that such deliveries are correctly certified will normally be a specialist task, but where delegated to the Clerk of Works the A/EA should be consulted in case of any doubt.

Sampling and testing: sampling of compost mixes for approval or test as directed by the A/EA should be arranged, and the results of the tests arranged by the contractor forwarded.

Replacements by the client: the conditions in the specification relating to relocation of plants, temperature, lack of water or unauthorised watering, change in lighting and effect of toxic chemicals that could place responsibility for replacement on the client/user should be noted. Any such problem arising should be reported to the A/EA and a record kept in the site diary.

Use of peat-based products: the use of peat is increasingly being criticised because of the environmental impact of extraction. Where specification excludes peat products, delivery should be verified for compliance.

Pesticides: similar concerns are raised over the use of certain pesticides, nitrate fertilisers, etc. The A/EA should advise on any environmental requirements. Unused chemicals and their containers are likely to constitute 'special waste' for disposal.

Maintenance: the A/EA should be able to supply a copy of the maintenance programme and instructions on any delegated duty. It is common for the landscape contractor to maintain the planting for a defined period. Liaison arrangements with the client/user for contractor's access after handover should be confirmed.

Notice to A/EA: where specified, the Clerk of Works should ensure that sufficient notice is given to the A/EA of the starting of specified operations.

Cleanliness: the site is to be left clean on completion to the A/EA's satisfaction.

Statutory requirements and technical standards

Publications to which reference may be necessary include:

BS 8000-0:2014 *Workmanship on Construction Sites. Introduction and General Principles*
BS 8000-13:1989 *Workmanship on Building Sites. Code of Practice for Above Ground Drainage and Sanitary Appliances.*
BS 8000-15:1990 *Workmanship on Building Sites. Code of Practice for Hot and Cold Water Services (Domestic Scale).*

7.11 BUILDING FABRIC SUNDRIES

Contract requirements

7.11.1 Duties

Part of the normal duty of the Clerk of Works is to verify for conformity with the specifications and drawings and, in the case of ironmongery, any schedules produced. It is important that cross-checks with a door schedule (if produced) are carried out, as external door manufacturers may have their own locking provisions or isolated trims to be included. The A/EA may utilise the services of a specialist to produce an ironmongery schedule.

Quality control: skirtings/architraves/trims/sundries

7.11.2 Quality assurance

The specification should be verified for applicable QA schemes and the A/EA's instructions obtained on any duties arising. For skirtings and architraves, reference is made to the TRADA scheme under Trussed rafters (see 7.5.14).

7.11.3 Skirtings and architraves

The specification should be verified for requirements to provide samples beforehand, and when delivered to site off-loaded materials should be checked for any splits, shakes, excessive knots or ribbed planed timber. Any disputes for rejection should be immediately referred to the A/EA.

Softwood material due to be painted should be treated for knotting, and stopped and primed prior to cutting into the required lengths for fixing. Softwood or hardwood due to be varnished should be carefully checked for imperfections prior to the finish being applied. Where hardwood skirtings and architraves are fixed via pelleting, the Clerk of Works should verify that the pellets are as close a match as possible to the surrounding timber and, when glued in, that they run with the grain of the surrounding material, as post-varnishing will highlight cross-graining.

Moisture content at the time of fixing should be between 8 and 12 per cent.

7.11.4 Door, Doorsets and Trims

For doors, doorsets, timber trims, quadrants and glazing beads, the specification should be checked for sample requirements beforehand. Delivery and finishes inspection should be carried out as for skirtings and architraves. (See 7.11.3.)

Internal Fire doors and doorsets ideally should be part of an approved fire safety scheme, similar to the British Woodworking Federation's "Certifire" scheme, so the Clerk of Works should verify that the correct FD30 or FD60 is fitted to the manufacturer's instructions, and the fire safety scheme label is visible. Gaps round edges of fire doors to frames should be as doors manufacturer instructions. Fire doors should be fitted by a competent person.

Softwood trims are normally fixed with galvanised or sheridised nails and stopped prior to decoration. The Clerk of Works should be mindful of excessive splits and refer any disputes to the A/EA.

Metal trims around door or window frames, where not part of that particular trade package, should be non-ferrous, straight and true. Where powder-coated to match existing trims, the Clerk of Works should be vigilant to check for a true colour match and report any non-compliance to the A/EA.

7.11.5 Sundry items

Often included in sundry items, but too numerous to have their own relevant sections, are:

- plywood shelves
- medium density fibreboard (MDF) window boards or boxings
- visible timber packers to posts, linings or frames.

Quality control: ironmongery

7.11.6 Quality assurance

The specification and schedule should be verified for applicable QA schemes and the A/EA's instructions obtained before proceeding further.

7.11.7 Manufacturer's instructions

Requirements in the specification to fix in accordance with the manufacturer's instructions or recommendations should be identified, copies of these obtained and compliance checked. It is often better for the site inspector to obtain a copy of the manufacturer's brochure, for readily identifiable reference.

7.11.8 Samples

The specification should be verified for requirements to submit samples, for example where powder-coated, and these must be submitted in good time for the A/EA's approval. Approved samples should be kept on site to check against deliveries.

7.11.9 Special points to watch

Lubrication: as ironmongery generally has several moving parts, it must be lubricated before installation, and, in any case, the manufacturer's instructions must be verified. A sealed case should never be broken for lubrication purposes, as this might invalidate any guarantees.

Door signage: the A/EA's agreement should be sought for heights and positions of door signage. Consistency throughout gives a neater and more professional appearance.

Handle fixing: before the tradespeople fix handles to internal timber doors, the Clerk of Works must make sure that the screws used are long enough to do the job. Far too often shorter screws pull out with misuse, in many instances taking a piece of flaxboard core with them. For heavily used doors, through-bolts might be a better option. In cases of doubt, the A/EA should be consulted.

Completion: at completion, the Clerk of Works should verify that Intumescent strips and smoke seals in fire door frames are present. The Clerk of Works should also verify that the ironmongery has been adjusted and lubricated if necessary and ensure that any protective trims are removed. It is also worth checking to see if the appropriate fire-retardant backing is present, as appropriate behind locks and hinges.

Statutory requirements and technical standards

Publications to which site reference may be necessary include:

Doors, doorsets, skirtings, architraves, trims and sundry items

BS 8214:2016 – *Timber-Based Fire Door Assemblies – Code of Practice*
BS 1186-3:1990 *Timber For and Workmanship in Joinery. Specification for Wood Trim and its Fixing*
BS 5499-4:2013 *Safety Signs, Including Fire Safety Signs. Code of Practice for Escape Route Signing*
BS ISO 3864-1:2011 – *Graphical Symbols. Safety Colours and Safety Signs. Design Principles for Safety Signs and Safety Markings*
BS ISO 3864-3:2012 *Graphical Symbols. Safety Colours and Safety Signs. Design Principles for Graphical Symbols for use in Safety Signs*
P20 *Unframed Isolated Trims/Skirtings/Sundry Items*

Requirements

BS 8000-0:2014 – *Workmanship on Construction Sites. Introduction and General Principles*
BS 8000-5:1990 *Workmanship on Building Sites. Code of Practice for Carpentry, Joinery and General Fixings*
BS EN 314-2:1993 *Plywood. Bonding Quality. Requirements*
BS EN 635-1:1995 *Plywood. Classification by Surface Appearance. General.*
BS EN 635-2:1995 *Plywood. Classification by Surface Appearance. Hardwood*
BS EN 635-3:1995 *Plywood. Classification by Surface Appearance. Softwood*
BS EN 635-5:1999 *Plywood. Classification by Surface Appearance. Methods for Measuring and Expressing Characteristics and Defects*
BS ISO 20712-1:2008 *Water Safety Signs and Beach Safety Flags. Specifications for Water Safety Signs used in Workplaces and Public Areas*

Ironmongery

BS 476-22:1987 *Fire Tests on Building Materials and Structures. Methods for Determination of the Fire Resistance of Non-Loadbearing Elements of Construction*
BS 8000-0:2014 *Workmanship on Construction Sites. Introduction and General Principles*
BS 8000-5:1990 *Workmanship on Building Sites. Code of Practice for Carpentry, Joinery and General Fixings*
BS EN 12209:2016 *Building Hardware. Mechanically Operated Locks and Locking Plates. Requirements and Test Methods*

7.12 PAVING, EXTERNAL PLANTING, FENCING AND SITE FURNITURE

Contract requirements

7.12.1 Duties

It is part of the Clerk of Works' normal duties to verify conformity with the specifications and drawings. Paving, planting and sports surfaces may require landscape or other specialist supervision, which the A/EA will arrange, but for small-scale work this duty may fall to the Clerk of Works. The A/EA will provide instructions on these matters and contact points for specialist advice.

7.12.2 A/EA's approval

Matters that may require prior A/EA's approval under the specification are set out in the respective work sections below.

Quality control

7.12.3 Quality assurance

The specification should be verified for applicable QA schemes and the A/EA's instructions obtained on any duties arising. In this workgroup reference is made to BSI-registered firms and BSI Kitemarks.

7.12.4 Paving

Setting out: the setting out of road lines, kerb lines, radii and level datum points should be verified. Levels and grades should be rechecked as work proceeds.

Materials: delivery documents should be verified for evidence of compliance with specification requirements. The contractor should be asked to provide additional documentation as necessary. The A/EA should be consulted in case of any doubt. In cases where the contractor offers materials as equivalent to those specified, this should be referred to the A/EA for approval or otherwise. The following requirements should be noted:

- **frost-resistant materials:** the contractor should be asked to provide evidence (in the form of test results) that the materials provided are non-frost susceptible, as defined in TRL SR 829; the A/EA should be consulted as necessary
- **cement:** should be obtained from a BSI-registered firm
- **steel:** dowel bars, mesh reinforcement and tie bars are to be obtained from a CARES licensee
- **thermoplastic paint:** to be supplied by a BSI Kitemarked licensee.

7.12.5 A/EA's approval

The following matters may need contractual approval as specified, and any delegations by the A/EA to the Clerk of Works should be established:

- **waterproofing compounds:** material to seal edges of cement-bound sub-bases and road bases to be approved
- **air-entraining agents:** to be approved (e.g. for in situ concrete)
- **bond-breaking agents:** to be approved (e.g. for in situ concrete)
- **blinding materials:** alternatives to sand for curing concrete, if proposed, require approval
- **surface finishes:** pedestrian pavings may be finished to the A/EA's requirements
- **bitumen:** bitumen binders are generally to be penetration grade unless the A/EA authorises the use of cutback
- **reflecting studs:** the A/EA is to approve the supplier
- **adhesion agents** (to improve adhesion between cutback bitumen if approved and uncoated chippings): to be approved, in conformity with *TRL Road Note 39*, 5th edition, 2002.

7.12.6 Testing

Where sampling and testing are to be carried out on paving materials, the A/EA will approve the test laboratory for the purpose. Test results are to be forwarded to the A/EA. On large-scale works, site testing laboratories may be set up, in which case the Clerk of Works' particular duties will be as directed by the A/EA. On small works, the Clerk of Works may be required to carry out sieve analysis checks on grading of aggregates. Where materials are mixed off site, the A/EA will arrange for any off-site inspection.

7.12.7 Workmanship

The following contractual requirements should be noted, and documentation or the A/EA's instructions obtained as necessary.

Type 2 sub-base: requirements for verifying compaction with correct moisture content should be agreed.

Concrete bases: requirements for verifying compaction to required average dry density should be agreed.

Trial concrete mixes: the A/EA's instructions on approval of trial mixes should be obtained.

Air entrainment: the Clerk of Works should approve the pressure-type meter to be used, witness tests, verify calibration of the meter and monitor compliance with the specification in accordance with BS EN 12350-7:2009.

Coated macadam: workmanship is to be in accordance with BS EN 13108-1, BS EN 13108-7 and BS 434-2.

Hot-rolled asphalt: workmanship is to be in accordance with BS EN 13108-4, by machine unless otherwise instructed (the A/EA's requirement should be established).

Mastic asphalt: workmanship is to be in accordance with BS EN 13108. The Clerk of Works should verify that the contractor has a suitable thermometer on site and uses it to conform to specified temperature limits for laying.

Dense tar: workmanship is to be in accordance with BS 5273.

Surface dressing: workmanship is to be in accordance with *TRL Road Note 39*, 5th edition, 2002. The A/EA's instructions should be obtained on whether to use section 7 or 8.

Slurry sealing: workmanship is to be in accordance with BS 434.

Tack coats: to be laid in accordance with BS 434-2.

Perforated concrete units: a copy of the manufacturer's instructions for laying should be obtained and compliance monitored.

Interlocking blocks: workmanship is to be in accordance with BS 7533. Manufacturer's instructions for any special requirements for the laying and protection of blocks during vibration should be obtained and compliance monitored. If samples are required to be approved by the A/EA, the A/EA must be given notice of readiness for inspection in good time.

Sand: sand for brushing into joints must be completely dry otherwise gaps may not be completely filled. Project specification for oven-dried sand or similar should be verified and compliance monitored.

7.12.8 Special surfacing and pavings for sport

This work will usually be carried out by a specialist contractor and may involve landscape or other specialist supervision. Such visits (contractor and supervisor) should be recorded in the site diary. The A/EA's instructions should be obtained on the Clerk of Works' particular duties.

7.12.9 Seeding and turfing

This work will usually be carried out by a landscape contractor and may involve supervision by a consultant appointed by the client or A/EA. Such visits should be recorded in the site diary. The A/EA's instructions on the Clerk of Works' particular duties should be obtained. Certification of materials delivered to the site should be verified and any discrepancies reported to the A/EA. Prior notice to the A/EA may be required for setting out, application of weedkillers and fertilisers, seeding, turfing, initial cutting and for maintenance visits by the specialist contractor. Such notice should be forwarded to the A/EA without delay.

7.12.10 A/EA's approval or decision

The following may require contractual approval or decision by the A/EA, and agreement of any delegations to the Clerk of Works:

- **samples of seed and turf:** for approval
- **equivalent manufacturer:** for approval, where offered
- **worm- and weedkillers:** approval of manufacturer; to be applied when directed by the A/EA
- **weather and site conditions:** to be agreed as suitable for work to proceed
- **damage:** reinstatement to the A/EA's satisfaction of grassed areas
- **turf stacking:** approval where period exceeds seven days, and on completion
- **cutting grass:** prior consent is required for the initial cut
- **re-seeding:** where growth is unacceptable to the A/EA
- **protective fencing:** to be removed when directed by the A/EA
- **rolling:** to be directed by the A/EA
- **top dressing:** to be directed by the A/EA
- **Tree Preservation Orders:** the Clerk of Works must verify with the A/EA that none exist on the site and obtain directions for protection if required by the A/EA.

7.12.11 Planting

This work will usually be carried out by a specialist contractor and may involve supervision by a consultant appointed by the client or A/EA. Such visits should be recorded in the site diary. The A/EA's instructions on the Clerk of Works' particular duties should be obtained. Certification of materials delivered to the site should be verified, and any discrepancies reported to the A/EA. Prior notice to the A/EA by the contractor may be required for setting out, weedkilling, fertilising, delivery, plant and maintenance visits.

7.12.12 A/EA's approval and decisions

The following may require contractual approval or decision by the A/EA. Any delegations to the Clerk of Works must be agreed.

- **Certification:** the contractor must provide certification of materials and compliance with specification where required by the A/EA.

- **Alternatives:** the contractor may propose alternative plants or equivalent manufacturers for the A/EA's approval.
- **Inspection:** the A/EA is to inspect and approve plants in the nursery prior to lifting.
- **Compaction:** the A/EA is to decide when soil has become compacted and requires to be broken up.
- **Hedge plants:** the A/EA is to instruct if leaders are to be cut back.
- **Watering:** the A/EA is to decide whether or not to water-in plants to saturation.
- **Sub-soil:** the A/EA is to direct disposal of excavated sub-soil.
- **Stakes:** tree stakes should be sawn off below the first branch, unless the A/EA instructs otherwise.
- **Stumps:** to be disposed of as directed.
- **Tree surgery:** the A/EA's prior permission may be required for use of retained trees as anchors for winching, use of artificial drying before application of fungicides, use of filler in splits, removal of major branches and severing of roots. Any defects or weaknesses not covered by the specification should be reported to the A/EA.
- **Burning:** ash is to be disposed of as directed. However, site fires may not be permissible and this should be verified with the A/EA.
- **Fencing:** the A/EA is to direct when to remove protective fencing.
- **Chemicals:** prior approval of the A/EA may be required for application of chemicals to control weeds. The Health and Safety Plan should be consulted for method statements.

7.12.13 Fencing

Fence erection: this may be specified to be carried out by a BSI-registered firm or by a specialist subcontractor, if not by the main contractor. Generally, fence erection should be carried out in accordance with the appropriate part of BS 1722.

Preservative: any proposal by the contractor to provide an equivalent to the specified treatment should be referred to the A/EA for approval or otherwise.

Statutory requirements and technical standards

Publications to which reference may be necessary include:

Roads and pavings

BS 434-2:2006 *Bitumen Road Emulsions. Code of Practice for the Use of Cationic Bitumen Emulsions on Roads and Other Paved Areas*
BS 5273:1975 *Specification. Dense Tar Surfacing for Roads and Other Paved Areas*
BS 7533 *Pavements Constructed with Clay, Natural Stone or Concrete Pavers*
BS EN 13108-1:2016 *Bituminous Mixtures. Material Specifications. Asphalt Concrete*
BS EN 13108-7:2016 *Bituminous Mixtures. Material Specifications. Porous Asphalt*
BS EN 12350-7:2009 *Testing Fresh Concrete. Air Content. Pressure Methods*

Transport Research Laboratory (TRL) SR 829 *Specification for the TRRL Frost-Heave Test, 1984*
TRL *Road Note 39: Design Guide for Road Surface Dressing, 5th edition, 2002*

Landscape

BS 1722 *Parts 1 to 14: Fences*
BS 3882:2015 *Specification for Top Soil*
BS 3969:1998 +A1:2013 *Recommendations for Turf for General Purposes*
BS 3998: 2010 *Tree Work. Recommendations*
BS 4428:1989 *Code of Practice for General Landscape Operations (Excluding Hard Surfaces)*
BS 4043:1989 *Recommendations for Transplanting Root-Balled Trees*

7.13 DISPOSAL SYSTEMS

Contract requirements

7.13.1 Duties

Conformity with the specifications and drawings, and with the Building Regulations, should be verified. In addition, the Clerk of Works should ensure that the drainage system and appliances fitted to it are operational. Guidance on inspection is given in BS 8000-16. Guidance on the responsibilities of the A/EA arising in the specifications that may be delegated to the Clerk of Works is included in the duties listed in the remainder of this section.

Quality control: above-ground drainage and sanitary appliances

7.13.2 Quality assurance

The specification should be verified against any applicable product QA schemes, and the A/EA should be consulted about duties arising. For drainage work, use is made of BSI Kitemarked products and Agrément certification.

7.13.3 Special points to watch

Internal pipe work: where not shown on drawings, the line and grade of pipe runs should be agreed, but the maximum branch lengths allowed by Building Regulations should always be borne in mind.

Discharge errors: the Clerk of Works should listen for traps being syphoned when two or more adjoining fittings are being discharged. Back-fall errors are basic but are still a regular item found on snag lists.

Overflow connections: when performance testing sanitary fittings, the Clerk of Works must ensure that the overflow connections are checked for leaks in addition to their full operation.

Water supply: it is not uncommon for hot taps to be fitted to the cold water supply in error, and the cold tap to the hot water supply. To ensure that no accidents occur, the correct orientation of taps should be checked. The cold tap should be on the right-hand side of the basin (a convention helpful to blind and partially sighted people).

Testing internal pipe work: pressure must be released from the end furthest away from the pressure gauge to prove that the whole pipe has been under test.

Fire protection: any plastic drainage pipe over 50mm in diameter passing from one fire zone into another must have a fire sleeve or 1m of cast iron drain pipe either side of zone barrier to comply with Part B of the Building Regulations.

Air admittance valves (AAVs): where an internal AAV is fitted to the top of a stub stack, the valve must be mounted above the height of the highest sanitary appliance fitting overflow point. AAVs should not be used at the head of a drain.

Quality control: below-ground drainage

7.13.4 Setting out

External drainage: the Clerk of Works should verify that invert levels are the point of discharge of new drainage against the level shown on the drawings. The correct setting out and depth relative to building lines, roads, paths and other service runs must be confirmed. It is important that manholes and inspection chambers are correctly sited to avoid acute bends at branch connections. The A/EA's agreement to the use of a laser instrument instead of sight rails must be obtained if proposed. The Clerk of Works should confirm that the contractor sets up the lasers accurately.

7.13.5 Safety

The Health and Safety Plan should be verified for any special provisions with regard to safe working in trenches, etc. Any potentially dangerous situations should immediately be brought to the contractor's attention. The health and safety coordinator and A/EA should also be informed as necessary.

7.13.6 Special points to watch

Falls: falls to required levels should never be achieved by supporting pipe work on bricks, stones or any hard objects. The supporting medium should be consistent in support.

Rocker pipes: unless otherwise specified, a short length of pipe (normally under 300mm) from the manhole (rocker pipe) should be included to allow for differential movement between pipe runs and manholes. The junction of this pipe with the adjoining pipe run should have a movement joint of approximately 20mm if pipes are being concreted for protection.

Pipe protection: the pipe runs must have protection from loads and traffic vibration appropriate to the depth of cover.

Pressure testing: the Clerk of Works must always ensure that the pressure is released from the end of the pipe work furthest away from the test gauge to make certain that the whole pipe has been under test. All other tests should conform to the specification and BS 8000-16. The requirement for profile testing must be kept in mind: in case of any doubt the A/EA should be consulted and advice sought, especially with regard to CCTV surveys.

7.13.7   Materials

Manufacturer's instructions: the specification should be verified for reference to these, copies obtained and compliance monitored. This applies particularly to proprietary pipe couplings.

Granular fill: testing must be arranged as specified or as directed by the A/EA.

Coarse sand: the source must be approved.

Backfill: the reuse of trench spoil must be approved where suitable and subject to test.

Statutory requirements and technical standards

Publications to which reference may be necessary include:

BS 416-1:1990 *Discharge and ventilating pipes and fittings, sand-cast or spun in cast iron. Specification for spigot and socket systems*

BS 416-2:1990 *Discharge and ventilating pipes and fittings, sand-cast or spun in cast iron. Specification for socketless systems*

BS 437:2008 *Specification for cast iron drain pipes, fittings and their joints for socketed and socketless systems*

BS 6465-1:2006+A1:2009 *Sanitary installations. Code of practice for the design of sanitary facilities and scales of provision of sanitary and associated appliances*

BS 6465-2:2017 *Sanitary installations. Space recommendations. Code of practice*

BS 8000-0:2014 *Workmanship on construction sites. Introduction and general principles*

BS 8000-13:1989 *Workmanship on building sites. Code of practice for above ground drainage and sanitary appliances*

BS 8000-14:1989 *Workmanship on building sites. Code of practice for below ground drainage*

BS 8000-16:1997 *Workmanship on building sites. Code of practice for sealing joints in buildings using sealants*

BS EN 877:1999 +A1:2006 *Cast iron pipes and fittings, their joints and accessories for the evacuation of water from buildings. Requirements, test methods and quality assurance*

7.14 PIPED SUPPLY SYSTEMS

Contract requirements

7.14.1 Duties

Conformity with the specifications and drawings should be confirmed. The many variable performance requirements specified for products to be installed require particular vigilance in checking the documentation and certification of deliveries, and conformity with appropriate standards, bye-laws and other statutory requirements. Competent personnel are obligatory for some tasks, and the Clerk of Works may be asked to confirm their qualifications.

The specification should be verified against applicable product QA schemes, and the A/EA's instruction obtained on any duties arising.

Quality control: water supply systems

7.14.2 External pipe work

Excavation for pipe work runs should be dealt with under BS 8000-0. The Clerk of Works should verify that correct setting out and depth relative to building lines, roads, paths and other service runs has been achieved.

7.14.3 Internal pipe work

Where not shown on drawings, the line and grading of pipe runs and points of connection to risers should be agreed. Air-locks must be avoided. Cold and hot water pipe runs should be coordinated for appearance in exposed sections, but avoiding heat transfer from hot to cold pipe work. Any cutting through the building structure not specifically provided for in the drawings requires prior approval and instruction on sleeving and fire-stopping. In case of doubt the A/EA should be consulted. The spacing of support brackets is to conform to the project specification.

7.14.4 Lead

Where capillary fittings are to be utilised, the Clerk of Works should verify that lead-free solder is used to make the joints on potable water systems. If integral solder ring fittings are to be used, they must be stamped as lead-free.

7.14.5 Compliance with the Water Supply Regulations

All water installations must be installed, tested and disinfected in accordance with the project specification, local water bye-laws (if applicable) and the Water Supply Regulations.

Copies of test and inspection certificates including initial pipe work pressure test, disinfection, initial and residual level results and final sampling after flushing, plus bacteriological tests, are required. There may also be a requirement for Legionella assessment.

7.14.6 Disinfection

Method statements should be obtained from the contractor, paying particular attention to the disposal of flushing and disinfecting water. The Clerk of Works should verify with the A/EA, Environment Agency or local authority that the contractor's proposals are acceptable.

Flushing and disinfection must be monitored to ensure that procedures are properly carried out and that required chlorine residuals are obtained. Water samples should be taken and despatched as directed by the A/EA, and bacteriological reports must be available for handover. If there is any question that water or distribution systems may have been contaminated, the A/EA must be informed accordingly.

Treated water: water supplies that receive additional treatment over and above initial disinfection must be clearly labelled and identified to avoid confusion with standard mains water; colour coding should conform to BS 1710. After disinfection, the pipe work must be flushed with the same type of water that it is designed to distribute.

7.14.7 Pressure testing

A sudden escape of water at high test pressures can cause considerable damage and may even cause personal injury. Particular attention should be paid to any temporary anchorages, supports and temporary connections, especially where existing premises are at risk from leakage. When testing mains water installations and fire mains the pressure might be particularly excessive, in which case clearing the area and placing danger signs may be in order.

7.14.8 Storage cisterns

Storage cisterns must be checked to ensure that they comply with water bye-laws, especially with regard to support, placing, valves, access for maintenance and supply pipe work.

Quality control: gas supplies

7.14.9 Generally

All gas pipe work and installations, including final connections, must be carried out by trade-approved personnel, i.e. Gas Safe-registered by the body that approves trades for different types of gas work. A Gas Safe-registered operative is only approved to work on certain types of gas installations. Operatives approved only for domestic installations are not qualified to work on commercial gas installations. All tradespeople should carry a card to say what type of work they are allowed to undertake.

All installations and materials should comply with the relevant sections from a wide range of statutory instruments and British Standards, such as BS 6891. Those applicable should be listed in the contract specification but an outline list is included at the end of this section (pp118–19) for reference.

7.14.10 External pipe work

Excavation for pipe work runs should be dealt with under BS 8000-0. The Clerk of Works should verify the correct setting out and depth relative to building lines, roads, paths and other service runs has been achieved. Service routes must be confirmed to be fully coordinated with other service routes, with particular attention paid to areas where services cross.

7.14.11 Internal pipe work

Where not shown on drawings, the line and grading of pipe runs must be agreed. Pipe runs should be coordinated for appearance in exposed sections. Where distribution pipes are routed through ceiling voids, floor voids or ducts they should be naturally ventilated. In the event that the pipe run crosses a fire barrier within a ceiling void, then each fire-stopped section must be separately ventilated. Pipes should not pass through areas that could be prejudicial to their safety without special fire safety provisions; these may include enclosing the pipe within a sleeve or duct, which should be ventilated at each end. Any cutting through the building structure not specifically provided for in the drawings requires prior approval and instruction on sleeving and fire-stopping. In case of doubt the A/EA should be consulted.

Regulation 18 of the Gas Safety Regulations 1984 gives detailed information with regard to enclosed pipes, protection against failure due to movement, shortest routes through walls and gas-tight sleeves.

7.14.12 Entry to buildings

Where services enter buildings, particular attention should be paid to sealing the point of entry to prevent seepage from possible leaks on external pipe work entering the building. In the event that a service crosses a cavity wall it should be sleeved to prevent seepage into the cavity. Plastic-type gas entry pipe work is permitted inside a building, but should not be exposed to sunlight.

7.14.13 Primary meter

The location of the meter is subject to agreement with the gas utility company. The meter enclosure, be it a purpose-built box or meter room, should be ventilated to the outside air.

7.14.14   Secondary meters

Any secondary meter serving a kitchen or other building or building compartment, which because of its use, may present a particular fire hazard, should preferably be housed in a room or chamber used for no other purpose, ventilated directly to the open air and fire-separated from the building by a construction having a fire resistance required by the Building Regulations.

7.14.15   Isolation

Each appliance should be provided with a valve or cock in an accessible location. In addition, where more than one appliance is situated in a self-contained area the installation pipe serving that area must also be fitted with a readily accessible isolating valve or cock. Non-domestic kitchens and boiler rooms, etc. should be fitted with an isolating valve or cock, which should preferably be located at the point of exit from the room. This valve or cock is for emergency and maintenance isolation only and should be suitably labelled.

7.14.16   Flexible tubes

Where flexible tubes are to be used for the connection of kitchen appliances these must comply with the requirements of BS 669-2 and BS 6173. For commercial and industrial works, refer to the Institution of Gas Engineers and Managers (IGEM) utilisation procedure IGE/UP/2.

7.14.17   Pressure testing

Testing of gas services should be undertaken in accordance with IGEM publications. Where existing services are to be altered or added to, a pressure test to ascertain the condition of the existing service should be undertaken before modifications begin.

7.14.18   Protective equipotential bonding

The Institution of Engineering and Technology (IET) Wiring Regulations require that main protective bonding conductors are installed connecting the main earthing terminal to gas and water metallic installation pipes. Where gas pipes serve more than one building, protective bonding conductors shall be installed in each building.

7.14.19   Flues

The Clerk of Works should verify that flues are complete, continuous and free from obstruction, with reference to relevant Building Regulations.

7.14.20 Ventilation for appliances

The provision of adequate air supply for open-flued appliances and ventilation for gas appliances in accordance with the Building and Gas Safety Regulations should be confirmed.

7.14.21 Maintenance access

It must be possible to gain access to an appliance for operation, maintenance and inspection.

7.14.22 Identification

Pipe work should be colour coded throughout its entire length in accordance with BS 1710.

Statutory requirements and technical standards

Publications to which reference may be necessary include:

General

BS 8000-0:2014 *Workmanship on Construction Sites. Introduction and General Principles*
BS 8000-1:1989 *Workmanship on Building Sites. Code of Practice for Excavation and Filling*

Water

BS 8558:2015 *Guide to the Design, Installation, Testing and Maintenance of Services Supplying Water for Domestic use within Buildings and their Curtilages. Complimentary Guidance to BS EN 806*
BS 8000-0:2014 *Workmanship on Construction Sites. Introduction and General Principles*
BS 8000-15:1990 *Workmanship on Building Sites. Code of Practice for Hot and Cold Water Services (domestic scale)*
BS EN 806 series. *Specifications for Installations Inside Buildings Conveying Water for Human Consumption*

Gas

BS 1710:2014 *Specification for Identification of Pipelines and Services*
BS 6172:2010+A1:2017 *Specification for Installation, Servicing and Maintenance of Domestic Gas Cooking Appliances (2nd and 3rd Family Gases)*
BS 6173:2009 *Specification for Installation of Gas-Fired Catering Appliances for use in all Types of Catering Establishments (2nd and 3rd Family Gases)*

> BS 669-2:1997 *Flexible Hoses, End Fittings and Sockets for Gas Burning Appliances. Specification for Corrugated Metallic Flexible Hoses, Covers, End Fittings and Sockets for Catering Appliances Burning 1st, 2nd and 3rd Family Gases*
> BS 6891:2015 *Installation of Low Pressure Gas Pipe Work of up to 35 mm (R11/4 1 and a quarter) in Domestic Premises (2nd Family Gas). SpecificationGas Safety (Installation and Use) Regulations 1984. SI 1984/1358*
> *Building Regulations 2000. SI 2000/2531*
> *Gas Safety Regulations 1972 SI 1972/1178*
> *Institute of Gas Engineers and Managers IGE/UP/2 Gas Installation Pipe Work, Boosters and Compressors on Industrial and Commercial Premises*
> *Water Regulations Advisory Scheme (WRAS) Water Fittings and Materials Directory. Available online at www.wras.co.uk/Directory/*

7.15 MECHANICAL HEATING, COOLING AND REFRIGERATION SYSTEMS

Contract requirements

7.15.1 Duties

Conformity with the specifications and drawings should be verified. The many variable performance requirements specified for products to be installed require particular vigilance in verifying the documentation and certification of deliveries, and conformity with appropriate standards, bye-laws and other statutory requirements. Competent personnel are obligatory for some tasks (electrical, water, gas and oil) and the Clerk of Works may be asked to confirm their qualifications or registration. In some areas of the site there may be a permit to work system in operation, and any non-compliance with this system should be reported directly to the A/EA.

The specification should be verified against product QA schemes and the A/EA's instruction obtained on any duties arising.

Section 7.16 below deals with ventilation and air conditioning systems.

Quality control: mechanical heating – sources of heat

7.15.2 General

This section deals with boiler plant and other sources of heat within a mechanical services installation. The setting-out details of boiler plant must be verified with the contractor, in particular that adequate space is available around plant for its assembly and maintenance. Plant should be examined on arrival to site, any damage noted and a check made that tappings are adequately protected and capped or plugged. If plant is to be craned into the building, verify that it is slung in a manner approved by the manufacturer so as not to impose undue stress or strain. Where particularly large plant is being installed it may be necessary for this to be positioned well in advance of the rest of the installation, and it may require greater attention to protection during the remainder of the construction period. Particular attention should be given to future maintenance with regard to how the plant can be safely accessed or dismantled for replacement.

The Clerk of Works should verify all sources of heat are sited so as to minimise the risk of heat damage to the building fabric, and that means of isolation are easily and safely accessible in an emergency. They should be mounted and secured strictly in accordance with the manufacturer's instructions. Additionally, installation should be by qualified personnel, especially for gas-fired and electrical equipment. Checks should be made that adequate air for combustion is available and that airways are not obstructed by other building works if the plant is operational (dust can block airways and/or damage gas burners, leaving them in an unsafe condition). All safety devices must be confirmed as operational and secure from unauthorised access before the plant is commissioned.

7.15.3 Builders' work

The Clerk of Works should verify preparatory builders' work associated with the installation of plant and equipment has been undertaken in accordance with the approved drawings. Further, the Clerk of Works should also verify what allowance has been made for the draining of the systems, and what permanent drainage is provided in the boiler room to the foul drain – chemical discharge directly into rivers will contravene legislation enforced by the Environment Agency.

7.15.4 Ventilation

There must be adequate ventilation to the boiler room/location, both for the provision of combustion and relief air, and for the dissipation of heat. Refer to IGEM publications, British Standards, Gas Safe publications, the Chartered Institution of Building Services Engineers (CIBSE) *Guide B*, etc. for guidance. Where mechanical ventilation is provided this should be interlocked with the boiler controls to shut down equipment.

7.15.5 Boiler mountings

The contract documents and any associated documents should be referred to for requirements for the installation of boiler mountings. All mountings must be easily reachable and accessible for maintenance. Floor mountings in basements should be on a concrete plinth to prevent burners and electrics being submerged in water.

7.15.6 Tools

Where required by the contract documents, maintenance and operating tools must be provided. Where these are to be hung on a rack or similarly provided in the boiler room, their location must be agreed with the contractor and training for their use may be required.

7.15.7 Solid fuel

Solid fuel boilers are seldom selected at present, as maintenance and attendance costs are now higher than for automated gas- or oil-fired boilers. They can be manually or automatically stoked, or fed with pulverised fuel. Greater space is required around and in particular in front of these boilers for maintenance, stoking and removal of ash and clinker.

7.15.8 Oil fuel

Oil fuels are supplied in varying classifications designated by their viscosity. The heavier oils will require preheating and pumping from the storage tank to the burner. All fuels are susceptible to thickening at lower temperatures, and even the lighter fuels may need to be trace heated where oil lines are susceptible to freezing conditions. Tanks should be positioned within a bund, to allow for containment of oil to prevent any escape into the drainage system, that is not less than 110 per cent of total tank capacity. Bunding tanks are now used and should be designed for catastrophic failure, and manufactured from rotationally moulded UV-stabilised medium-density polyethylene. Advice from the local environmental offices should be sought.

At the point of entry to the boiler room a fire valve will be installed to cut off the supply of oil in case of fire. This may be mechanically or electrically activated.

A means of containing an oil spillage and raising an alarm should be provided. This will generally be in the form of a sump with a level switch, which should shut down the plant and cut off the oil supply. This facility must be easily accessible and maintainable.

7.15.9 Gas

Any requirements for secondary metering, or for the provision of a test meter at the burner should be noted. This is sometimes required in larger installations. At the point of entry to the boiler room, a gas cock or a fire valve will normally be installed to cut off the supply of gas in a fire situation. This may be mechanically or electrically activated by a fire alarm or emergency stop button, with a reset control fitted within the boiler room and incorporated (if possible) on the main control panel.

7.15.10 Combination boilers

Combination boilers are fed direct from the mains water supply, and in addition to providing heating they also provide domestic hot water with no storage. These should be installed in accordance with the Gas Regulations, with reference to the manufacturer's instructions and the requirements of the Water Regulations, in particular regarding cross-contamination of supply.

7.15.11 Condensing or high efficiency boilers

Conventional boiler design seeks to avoid flue gas side corrosion by condensation, and as a consequence boiler efficiency is limited to about 85 per cent. By comparison, a condensing boiler seeks to recover the latent heat in the flue gases by returning the boiler water at a lower temperature; efficiency can then be increased to about 93 per cent. The flue gas condensate will be slightly acidic and the boiler will be designed to accommodate this. The condensate from the boiler flue will, however, need to be piped away to a foul drain in a material suitably resistant to the acidity of the condensate. In addition, because of the low flue temperature (65°C), a natural draught will not be present in the flue and mechanical ventilation will be necessary. It should be noted that as the boiler return water temperature is lower than in conventional design, the system mean water temperature will therefore also be lower. This will result in radiators and other emitters having to be larger when compared to conventional design.

7.15.12 Electrically heated boilers

This type of boiler is heated with single-, twin- or three-phase electrical heating elements installed in the water pressure unit, and is primarily for small domestic premises or where there is no possible storage space or gas supply available. A competent person should be qualified and registered to install this type of pressurised system.

7.15.13 Multi-boiler installations

The control requirements for sequencing or step control should be noted. Motorised isolating valves are provided on the return connection to each boiler and will be interlocked with the burner operation to each boiler. Particular attention should be paid to the piping up of the boilers. A loop header system is becoming the norm, as it is proving to be the most flexible. It is not unusual for a reverse return method of piping to be adopted.

7.15.14 Specialist inspections

Boiler plant, in particular steam raising plant, may require inspection by insurance, safety and building control officers. The Clerk of Works should note the requirements of the contract documents, discuss them with the A/EA and liaise with any specialists as necessary. It is likely that plant will have been pressure tested off site, in which case copies of the relevant test certification should be obtained from the contractor and securely retained, but there may be a requirement for site-assembled plant to be on-site tested. The Clerk of Works should liaise between the contractor and any specialists in this latter instance.

Any system operating at temperatures and pressures defined to be medium or high is required to be examined by a competent person before it is put into use. It is recommended that for any sealed system the project engineer is consulted as early as possible and that the contractor is made fully aware in good time if there are requirements for additional inspections or tests, or if the client and engineer wish to apply any higher standards. The following section provides further information and advice.

Quality control: heat distribution systems

7.15.15 General

This section deals with the distribution systems for transfer of the heating medium throughout an installation. These may be steam and condensate, or low-, medium- or high-temperature hot water. This section covers both external and internal installations, and pre-insulated pipe work.

7.15.16 Water/oil systems

Heat distribution systems generally use water as the transfer medium, but occasionally oil may be encountered. For oil systems the mechanical engineer should be consulted if insufficient information is contained in the specification for the Clerk of Works' duties to be carried out effectively.

7.15.17 Storage of pipe work and materials

Pipe work should be stored clear of the ground on racking, with ends capped and protected and, ideally, sheltered from rain. Tube ends should be cleaned of scale, paint, oil, etc., and any rusting should be removed and treated with a suitable primer before installation. Threaded and reamed iron pipes should also be reprimed before installing. Pipe work fittings should be sorted, binned and stored under cover.

Valves and cocks, particularly with flanged faces, should be plugged or plated to prevent the ingress of foreign objects and damage to seats and seals. Expansion devices should be similarly protected and retaining or sizing bars left in place until installed (refer to the manufacturer's instructions).

7.15.18 Setting out

External pipe work: excavation for pipe work runs should be dealt with under BS 8000-0. The correct setting out, grading, venting and draining of pipe work and depth relative to building lines, roads, paths and other service runs should be confirmed. Sand covering of pipes and identification tapes are placed in the top 400mm of the trench to protect the services.

Internal pipe work: where not shown on drawings, the line and grading of pipe runs and points of connection to risers must be agreed. To avoid air-locks, vents should be positioned at high points and pipe work should be completely drainable to an accessible safe point. Particular attention should be paid to pipe gradients on steam pipe work where the effective removal of condensate is important.

Pipe runs should be coordinated for appearance in exposed sections but avoiding heat transfer from hot to cold pipe work. Any cutting through the building structure not specifically provided for in the drawings requires prior approval and instruction on sleeving and fire-stopping. In case of doubt the A/EA should be consulted.

7.15.19 Installation

For all systems, the Clerk of Works should verify that as work proceeds it is installed to its coordinated location, is adequately supported and braced, and is laid evenly to avoid trapped pockets of air. Supports should allow for thermal expansion, preventing stress to pipe joints.

Air bottles and vents need to be installed where designed and/or where high spots cannot be avoided. Equally, drain cocks should be provided to allow full and effective removal of all liquid from the system by gravity.

Open ends of pipe work left during the progress of work should be closed temporarily with purpose-made metal, plastic or wood plugs, caps or blank metal flanges. Joints should not be made in the thickness of any wall, floor or ceiling, and pipe work passing through the building structure should be sleeved in a material similar to the pipe. Sleeves must be fire-stopped where passing through a fire barrier. Galvanised fittings are only to be used on galvanised pipe work.

Particular attention must be paid to the bracketing of pipe work as required by the contract documents, and any requirements for the guiding and anchoring of pipe work for the purpose of expansion or bellows should be noted.

The type and manufacture of valves must be as specified, and they must be accessible.

7.15.20 Thermal insulation

The requirements of the contract documents for the application of thermal insulation should be noted: many specifiers require that pipe work be painted prior to the application of insulation to eliminate damage to the primer during installation. Pressure testing should be undertaken before the application of insulation, and on pre-insulated pipe work joints should be left exposed for testing.

Unless using pre-lagged pipe, no insulation should be applied until all pressure testing and inspections have been completed. However, if circumstances dictate that insulation starts earlier, no joints, either welded or screwed, should be covered until testing has been completed.

Insulation material should be checked for conductivity, quality and wall thickness against the specification. Valves and flanges must be insulated if specified and the correct surface finish applied. The Clerk of Works should verify that circuit name, colour code for medium, and directional arrows are correctly applied.

Adequate identification banding must be applied in accordance with British Standards and contract documents.

7.15.21 Flushing and dosing

Method statements should be obtained from the contractor, paying particular attention to the disposal of flushing and dosing water. The Clerk of Works should verify with the A/EA, Environment Agency, local authority and the local water authority/company that the contractor's proposals are acceptable.

Flushing and dosing must be properly carried out, and required water conditions obtained. A negative or acidic reading of pH 7.1 with a low bacteria count will help to prevent noise and a build-up of oxides. Water samples should be taken and despatched as directed by the A/EA. Disposal of flushing and dosing water should be in accordance with any discharge consents.

7.15.22 Chemical cleaning

If chemical cleaning is required, a method statement should be obtained via the main contractor from a specialist chemical cleaning contractor. Cleaning follows on from the flushing out of the systems to remove loose scale and other deposits in the pre-commissioning stage. The pipe work is thoroughly cleaned of oils, greases, mill scale and other corrosion-forming deposits. Refer to *CIBSE Commissioning Code W* for procedures of flushing, chemical cleaning and passivating.

7.15.23 Inspection

Many water systems need to be inspected and tested in accordance with the Health and Safety at Work etc. Act 1974 and the Pressure Systems and Transportable Gas Containers Regulations 1989. If the system that is being installed does not have an open vent, the specification should be verified against inspection and testing requirements to meet the regulations (for example by a competent person) and the project engineer consulted if these are not specified.

The pressurisation tank should be checked to ensure that it has the correct maximum pressure test stamped on the top or side to meet the design specifications. Also, the size of vent pipe must increase as the distance runs, and it should be confirmed that the pressure relief valve spring is for 'high' pressure (this should always be one size larger than entering pipe to vent).

Normally, it will be the responsibility of the contractor to engage the services of a competent person to carry out these duties, and the Clerk of Works should refrain from carrying out the duties of a competent person without the written authority of the project manager or client. The contractor should be asked to confirm in writing the company being employed to carry out these duties, and if the Clerk of Works is not familiar with the name and reputation of the company, it is reasonable to request a list of similar works that they have undertaken recently.

If any cause for concern still lingers, the project engineer and A/EA should be consulted. The competent person must witness any large-scale pressure testing of the system, unless they have visited the site and agreed to specific inspection and testing arrangements.

7.15.24 Pressure testing

A sudden escape of water or air at high test pressures can cause considerable damage, and may even cause personal injury. Particular attention must be paid to any temporary anchorages, supports and temporary connections, especially where existing premises are at risk from leakage. Adequate precautions must be taken to prevent the deformation or distortion of expansion devices. When testing steam or high-pressure hot water mains installations, the pressure may be particularly excessive, in which case clearing the area and placing danger signs may be necessary.

Where pre-insulated external mains are to be pressure tested, the test may be for a 24-hour duration and may also require monitoring of ambient air temperatures by chart recorder. Refer to the contract documents for further clarification. A check should always be made that pressure is present at the furthest point in pipe runs.

7.15.25 Certificates

Copies of all inspection and test certificates, including welder's assessment certificates where necessary, must be available for inclusion in the handover documents.

Quality control: refrigeration systems

7.15.26 Application

This section covers all refrigeration plant, refrigerant and chilled water pipe work.

7.15.27 Safety and environment

Although mainly aimed at refrigeration plant for air conditioning systems, the safety procedures also apply to split refrigeration installations for food storage or beer cellars.

Most refrigerant gases are hazardous to health, and extensive safety procedures must be followed when charging and discharging refrigerant systems. BS EN 378 contains detailed requirements for their safe design, construction and installation. Many refrigerants are also environmentally destructive and precautions must be taken to prevent release.

Hazard warnings are required to be displayed for semi-sealed systems, and gas detectors with audible alarms installed where natural ventilation of the room is not available. The Clerk of Works should ensure, in conjunction with the health and safety coordinator, that the Health and Safety Plan and the Health and Safety File at handover contain safety procedures.

7.15.28   Storage of pipe fittings and valves

Pipe work fittings should be sorted, binned or bagged and stored under cover. Valves and cocks, particularly those with flanged faces, should be plugged or plated to prevent the ingress of foreign objects and damage to seats and seals. Expansion devices should be similarly protected, and retaining or sizing bars left in place until installed (refer to the manufacturer's instructions).

7.15.29   Setting out

External pipe work: excavation for pipe work runs should be dealt with under BS 8000-0. The correct setting out, grading, venting and draining of pipe work and depth relative to building lines, roads, paths and other service runs must be confirmed.

Internal pipe work: where not shown on drawings, the line and grading of pipe runs and points of connection to risers should be agreed. Air-locks should be avoided, and pipe work must be completely drainable. Particular attention should be paid to pipe gradients. Pipe runs should be coordinated for appearance in exposed sections and services routed so as to avoid unwanted heat gain. Any specified noise attenuation must be incorporated and installed correctly.

7.15.30   Installation

Open ends of pipe work left during the progress of work should be closed temporarily with purpose-made metal, plastic or wood plugs, caps or blank metal flanges. Joints should not be made in the thickness of any wall, floor or ceiling, and pipe work passing through the building structure should be sleeved in a material similar to the pipe. Any cutting through the building structure not specifically provided for in the drawings requires prior approval. Where there is any doubt the A/EA should be consulted. Sleeves must be fire-stopped where passing through a fire barrier. Only galvanised fittings may be used on galvanised pipe work. Overflows will need a 1.5-degree slope to allow for flow, and a fire rating when passing through fire barriers.

Particular attention should be paid to the bracketing of pipe work as required by the contract documents, and any requirements for the guiding and anchoring of pipe work for the purpose of expansion should be noted.

7.15.31 Room air conditioners

These may be of the following types:

- single packaged, through wall or window
- ducted from main air conditioning plant
- split type with remote condenser unit
- split type for connection to remote central cooling plant.

They may also include, in certain circumstances, integral air heaters or humidifiers, with a flow and return to retain the altered environment.

7.15.32 Central plant cooling

Plant may be of the direct expansion (DX) type or chilled water coils connected to remote chiller plant. External units need to be positioned to prevent leaves and dirt clogging the exchanger and to allow easy access for maintenance.

7.15.33 Central plant cooling, heating and humidity

This plant would be either delivered complete or assembled on site, and consists of a cooling refrigerated heat exchanger, a heating coil, humidifier and filters connected to a circulation fan. This type of unit can recirculate or allow fresh air into rooms at a controlled temperature. Overflows can become blocked and provision should be made to prevent damage to equipment.

7.15.34 Chilled water plant

It is likely that a specialist engineer may be engaged to comment upon the installation of the plant and to oversee and witness the commissioning of the plant. In this event, the Clerk of Works should liaise through the A/EA with the engineer as required.

7.15.35 Refrigerant pipe work

Refrigerant pipe work may be of copper or steel. Pipe work of either material should be stored clear of the ground in clean and dry conditions, with ends sealed until required for installation. Pipe work should be supported as stated in the contract documents, and precautions taken to prevent cold bridging. Provisions should also be made to accommodate thermal expansion and contraction.

Copper pipe work: flared and brazed joints are the norm, although compression joints may be utilised for connections to pressure gauges and items of equipment. Brazing should be carried out in accordance with Heating and Ventilating Contractors'

Association (HVCA) codes of practice and the relevant British Standards. The Clerk of Works should verify that a scheme of approval for welders is instigated, and that welders achieve the necessary standard of competency.

Steel pipe work: welded or flanged joints are the norm, although screwed joints may be used on connections to equipment. Welding should be carried out by suitably qualified and certified operatives in accordance with HVCA codes of practice and BS 2971.

7.15.36 Chilled water pipe work

Chilled water pipe work will generally be of black mild steel, except for open condenser water systems, air washer, drain, vent and overflow pipe work, which will generally be of galvanised steel. Tube ends should be cleaned of scale, paint, oil, etc., and any rusting should be removed and treated with a suitable primer before installation. Galvanised fittings are only to be used on galvanised pipe work.

7.15.37 Thermal insulation

The requirements of the contract documents for the application of thermal insulation should be noted; many specifiers require that pipe work be painted prior to the application of insulation. Pressure testing should be undertaken before the application of insulation, and on pre-insulated pipe work joints should be left exposed for testing.

Chilled water and refrigerant pipe work insulation will normally incorporate a vapour barrier, and it is of the utmost importance that this is not penetrated by brackets or supports. Where the insulation is to be over-clad with hammer-clad metal, fixings must not be permitted to penetrate the vapour barrier. Adequate identification banding must be applied in accordance with the British Standards and contract documents.

7.15.38 Flushing and dosing

Method statements should be obtained from the contractor, with particular attention paid to the disposal of flushing and dosing water. The Clerk of Works should verify with the A/EA, Environment Agency and the local water authority that the contractor's proposals are acceptable.

Flushing and dosing must be properly carried out, and required water conditions must be obtained. Water samples should be taken and despatched as directed by the A/EA.

7.15.39 Catering

Refrigeration for catering purposes requires specific temperature ranges for different products. All equipment provided must be labelled for contents and be correct to specification and current legislation. Walk-in freezers should have external ventilation and safety devices to prevent the door from accidentally closing.

7.15.40 Performance testing

Performance testing using simulated loads to the rating detailed in the specification must be monitored during setting up and testing. All test instruments must be calibrated by an accredited agency and have current certificates of calibration.

Method statements should generally be produced by the contractor for charging of any system and for all testing.

Statutory requirements and technical standards

Publications to which reference may be necessary include:

BS 779:1989 *Specification for Cast iron Boilers for Central Heating and Indirect Hot Water Supply (Rated Output 44 kW and Above) [Partially replaced by: BS EN 303-1:1999 Heating Boilers. Heating Boilers with Forced Draught Burners. Terminology, General Requirements, Testing and Marking, and BS EN 303-4:1999 Heating Boilers. Heating Boilers with Forced Draught Burners. Heating Boilers with Forced Draught Burners. Special Requirements for Boilers with Forced Draught Oil Burners with Outputs up to 70kW and a Maximum Operating Pressure of 3 bar. Terminology, Special Requirements, Testing and Marking]*

BS 855:1990 *Specification for Welded Steel Boilers for Central Heating and Indirect Hot Water Supply (Rated Output 44kW to 3MW) [Partially replaced by BS EN 12797:2000 Brazing. Destructive Tests of Brazed Joints]*

BS 1113:1999 *Specification for Design and Manufacture of Water-Tube Steam Generating Plant (Including Superheaters, Reheaters and Steel Tube Economisers) [Partially replaced by BS EN 12952 Parts 1 to 7, 10 and 11 Water-tube Boilers and Auxiliary Installations]*

BS 1894:1992 *Specification for Design and Manufacture of Electric Boilers of Welded Construction [Partially replaced by BS EN 12953-8:2001 Shell Boilers. Requirements for Safeguards Against Excessive Pressure]*

BS 2790: *Specification for Design and Manufacture of Shell Boilers of Welded Construction [Partially replaced by BS EN 12953 parts 1 to 6 and part 8 Shell Boilers]*

BS 2971:1991: *Specification for Class II Arc Welding of Carbon Steel Pipe Work for Carrying Fluids*

BS 8000-0:2014 *Workmanship on Construction Sites. Introduction and General Principles*

BS 8000-1:1989 *Workmanship on Building Sites. Code of Practice for Excavation and Filling*

BS EN 301-1:2017 *Heating Boilers. Heating Boilers with Forced Draught Burners. Terminology, General Requirements, Testing and Marking, and BS EN 303-4:1999 Heating Boilers. Heating Boilers with Forced Draught Burners. Heating Boilers with Forced Draught Burners. Special Requirements for Boilers with Forced Draught Oil Burners with Outputs up to 70kW and a Maximum Operating Pressure of 3 Bar. Terminology, Special Requirements, Testing and Marking*

> BS EN 378 series *Specification for Refrigerating Systems and Heat Pumps*
> BS EN 12799:2000 *Brazing. Non-Destructive Examination of Brazed Joints*
> BS EN 13134:2000 *Brazing. Procedure Approval*
> BS EN 14324:2004 *Brazing. Guidance on the Application of Brazed Joints*
> BS EN ISO 13585:2012 *Brazing. Qualification Test of Brazers and Brazing Operators*
> *Health and Safety at Work etc. Act 1974. Chapter 37*
> *PD 5500:2015+A2:2016 Specification for Unfired Fusion Welded Pressure Vessels*
> *Pressure Systems and Transportable Gas Containers Regulations 1989 SI 1989/2169*
> *CIBSE Commissioning Code W, Water Distribution Systems*
>
> HVCA publications
>
> *TR3 – Jointing of Copper and its Alloys*
> *TR5 – Welding of Carbon Steel Pipework*
> *TR6 – Guide to Good Practice for Site Pressure Testing of Pipework*

7.16 MECHANICAL VENTILATION AND AIR CONDITIONING SYSTEMS

Contract requirements

7.16.1 Duties

Conformity with specifications and drawings should be verified. The many variable performance requirements specified for products to be installed require particular vigilance in verifying the documentation and certification of deliveries and conformity with appropriate standards, bye-laws and other statutory requirements. Competent personnel are obligatory for some tasks, and the Clerk of Works may be asked to confirm their qualifications.

The specification should be verified for applicable product QA schemes and the A/EA's instruction obtained on any duties arising.

This section should be read in conjunction with the refrigeration systems section from 7.15.26 to 7.15.40.

Quality control

7.16.2 Design

These types of systems are always purpose-designed and manufactured, with the specialist contractor carrying out detail design to suit the performance requirements stipulated in the particular specification and the building as constructed. Any drawings produced should be submitted to the project engineer for approval before manufacture starts, and copies of approved drawings provided to the site inspector before installation starts.

7.16.3   Delivery, on-site storage and handling

Delivery of material to site should be monitored, and checks made to ensure that it is stored safely and weather-protected.

If the site is not ready to receive ductwork and plant, adequate facilities must be provided for storage. Ductwork should be stored clear of the ground and in such a manner that it is not subject to damage or deformation. Open ends of ductwork and plant should be protected against the ingress of debris during storage until they are installed; during installation any open ends should be similarly protected. Dampers and terminal devices should be stored away from the working area until they are installed.

Adequate measures should be in place for the handling and moving of ductwork and plant items from the place of off-loading/storage to the workplace, with due regard for bulky and heavy items.

7.16.4   Installation procedures

Ductwork should be verified to ensure that it is of the correct gauge as required by the contract documents. Where sheet metal ductwork is to be cut on site, the Clerk of Works should confirm that the correct tools are being used, that all raw or rough edges, both externally and internally, are removed, and that areas where galvanising has been destroyed are cleaned, prepared and painted with zinc-rich paint. Checks should be carried out to ensure that no use is made of plant or ductwork as access platforms by following trades to undertake their own work. Flexible and bendable ducts should be inspected to ensure that they are not deformed and are not bent to a radius outside the manufacturer's tolerances. The installation should be verified against the contractor's working drawings as it proceeds, with any deviations noted.

7.16.5   Air handling plant

Plant items should be examined and verified for compliance with the contract documents and approved manufacturer's drawings. Any deviations are to be noted in the site diary and the contractor and A/EA notified. When installed, the Clerk of Works must verify the plant has sufficient space around it for routine maintenance to be undertaken, that pipe work connected to the plant does not prevent the opening of access doors or the removal of heating or cooling coils, etc., and that air handling unit sections are properly assembled, access doors are properly seated and there are no air leaks on the plant. Safety guards should be fitted to all fan drives.

All heating and cooling coils must be examined for damage and, if necessary, the contractor should comb the fins of the coils.

Where cooling coils are fitted, adequate allowance must be made for the collection and removal of condensate caused by the dehumidification process. Similarly, where humidifiers are installed, provision should be made for the removal of excess water. Room cooler units mounted in ceiling voids and at floor level may present particular problems in this respect.

Adequate bypasses must be fitted to heating/cooling coils for flushing.

7.16.6 Fans

Fans must be properly mounted in accordance with the manufacturer's instructions, and adequate provision should be made for vibration isolation, by the use of anti-vibration mounts and flexible duct connections. The shaft and impeller assembly of all fans should be statically and dynamically balanced to BS ISO 21940-12:2016.

An equipotential bonding vibration loop should be provided across flexible duct connections.

7.16.7 Ductwork

The Clerk of Works should verify that ductwork is effectively supported as installation proceeds, and that insulation strips are incorporated into supports where required. After it has been levelled and straightened, checks for cleanliness should be witnessed before testing and commissioning starts.

Test procedures and results should be witnessed and calibration of test instruments and certificates of calibration verified. The specified test routine must be followed correctly, and any results that are out of specification referred to the project engineer. If a high-efficiency or a specific filter material has been called for, the Clerk of Works should verify that the appropriate filter has been installed and that the airflow direction is correct.

The contract requirements for the leakage testing of ductwork should be verified and witnessed as necessary.

7.16.8 Filter replacement

The requirements of the contract documents for the provision and replacement of filter media before handover should be noted. Generally, all filters should be clean at the time of handover.

7.16.9 Noise attenuators

The ends of attenuators are to be kept covered until they are installed. Before installation, they should be verified internally for any damage, and any non-compliant units rejected.

7.16.10 Instruments

The requirements of the contract documents with regard to the installation of instrumentation, thermometers and pressure gauges, etc. should be noted. These must be installed in accordance with the manufacturer's requirements in areas where they will provide a representative reading.

7.16.11 Thermal insulation

Where thermal insulation is to be applied to ventilation ductwork, the Clerk of Works should pay due regard to vapour barriers and ensure that duct access doors and test points are not obstructed or covered. Adequate provision must be made at points of support to prevent the piercing of vapour barriers and the compression of the insulation material. Support-piercing fasteners should be examined to ensure that they are adequately fixed to the ductwork surface with suitable adhesive. Problems are often encountered with 'self-adhesive stick pin' fasteners because oil residues on ductwork surfaces are not properly cleaned off prior to fixing. Identification triangles and lettering should be applied in accordance with the contract documents.

7.16.12 Dampers

Fire and regulating dampers should be installed with sufficient access for their maintenance, and sufficient and adequate access doors must be fitted in the ductwork. Fire dampers should be installed in proper mounting frames within the thickness of firebreak walls or floors. All dampers should be verified for operation in accordance with the CIBSE Commissioning Codes before the plant is put into operation.

The operation of any mechanical interlocks or connections with fire alarm interfaces should be examined during the verification of automatic controls.

7.16.13 Chillers

Where water treatment is required on chilled water systems or humidifiers, the Clerk of Works should verify that dosing is to correct levels, that any safety labelling required under the Control of Substances Hazardous to Health Regulations 2002 (COSHH) is in place, and that the Health and Safety Plan and File reflect safety procedures. Additionally, for evaporation cooling towers all required water tests and treatment, including Legionella or associated bacteriological tests, must be certified as having been complied with fully.

Some refrigerants are banned under the Montreal Protocol. Others that are permitted may still have a significant environmental impact if they are released into the environment, and stringent precautions should be in place. The specification should be verified for any specific requirements but, additionally, the Clerk of Works should be familiar with the refrigerants being proposed/used and relevant statutory requirements or environmental good practice.

7.16.14   Fume and industrial extraction

Generally, the procedures to be followed are those for all ductwork, but there is a need for closer inspection regarding quality standards. Environmental health, chemical safety and biological containment needs mean that any extraction system dealing with the removal from a work station of airborne pollutants must be constructed and installed to a high standard, and detailed records kept of the efficiency and performance achieved, as these are used as the benchmark for later maintenance inspections of the installation. Therefore, the system must be installed to the design and meet or exceed the specified performance criteria: any short-comings must be reported to the A/EA and project engineer.

The need for diligence and care in the monitoring and recording of an extraction system that may handle a hazardous substance cannot be overstated, as harm or injury to a person using the facility at a later date caused by the substance that it is designed to handle may result in legal action.

If the Clerk of Works is aware of the substance that the system is designed to handle, data on its level of hazard to health can be obtained from HSE Publication EH 40/2005 *Workplace Exposure Limits*.

Statutory requirements and technical standards

Publications to which reference may be necessary include:

BS ISO 21940-12:2016 *Mechanical Vibration. Rotor Balancing. Procedures and Tolerances for Rotors with Flexible Behaviour*
BS ISO 21940-11:2016 *Mechanical Vibration. Rotor Balancing. Procedures and Tolerances for Rotors with Rigid Behaviour*
DW/143 *Practical Guide to Ductwork Leakage Testing*
DW/144 *Specification for Sheet Metal Ductwork: Low, Medium and High Pressure/ Velocity Air Systems*
DW/154 *Specification for Plastics Ductwork*
DW/172 *Standard for Kitchen Ventilation Systems*
DW/191 *Guide to Good Practice – Glass Fibre Ductwork*
(DW leaflets are available from HVCA Publications: see address in Chapter 17.)
HSE EH 40/2005 *Workplace Exposure Limits*

7.17   ELECTRICITY SUPPLY, POWER AND LIGHTING SYSTEMS

Contract requirements

7.17.1   Duties

Conformity with the specifications and drawings should be verified. The many variable performance requirements specified for products to be installed require particular vigilance in checking the documentation and certification of deliveries and conformity with appropriate standards, bye-laws and other statutory requirements. Competent personnel are obligatory for some tasks, and the Clerk of Works may be asked to confirm their qualifications.

The specification should be verified against applicable product QA schemes and the A/EA's instruction obtained on any duties arising.

Quality control: high voltage – generation, supply and distribution

7.17.2 Definitions

Under BS 7671:2008+A3:2015 low voltage (LV) is defined as:

- not exceeding 1,000V AC or 1,500V DC between conductors
- 600V AC or 900V DC from any conductor to earth.

High voltage (HV) is therefore any voltage exceeding the above definition.

7.17.3 Authorised person

Any connection to or alteration of an HV system can only be carried out under the auspices of an authorised person (AP) as defined under the Health and Safety at Work etc. Act 1974, Electricity at Work Regulations 1989 and Electricity Supply Regulations 1988, and by people skilled and experienced in such work.

Reference should also be made to the preliminaries of the project specification and the Health and Safety Plan for the duties of APs. In association with the A/EA, the Clerk of Works should liaise between all parties to agree any demarcation and division of responsibilities.

No jointing or testing must proceed unless a method statement is available that has been verified and countersigned by a second AP.

7.17.4 Cable routes

In agreement with the project engineer, the Clerk of Works should confirm all cable routes and ensure that cable clearances have been gained. Particular attention should be paid to the safety of excavations and the trench bed before cables are laid.

Cables should be laid strictly in accordance with manufacturer's instructions, and all jointing carried out as detailed by the jointing system manufacturer. The ambient temperature should not fall outside the manufacturer's specified limits, and jointing must be carried out under clean and acceptable conditions, with staging to avoid contamination if the trench is wet or muddy, and weather protection available. Cable radii must be in accordance with the manufacturer's recommendations.

7.17.5 Cable identification

For connection into, or alteration of, an existing system the role of the Clerk of Works is strictly limited to witnessing installation procedures and testing. It is the responsibility of the AP to identify cables, and carry out all switching, earthing and proving dead. The AP should also be completely satisfied that connections have been made correctly and colour-true before cancelling the permit to work.

If permanent identification labels are not available it is suggested that suitable temporary labels identifying each switchgear circuit are fitted before the permit to work is issued to avoid possible confusion and mismatching of cable to switch, as this could have serious operational consequences.

Despite the above advice, the Clerk of Works is strongly advised to refrain from identifying existing cables or circuits and to leave this to the AP. Full details including times of issue and cancellation of permits to work should be recorded. A copy of a permit to work, up to and including cancellation, should be obtained.

7.17.6 Transformers

After receipt on site, it should be confirmed that the voltage ratio, vector group and winding type are as specified. The transformer should be installed into a bunded area of sufficient capacity to contain the entire oil/liquid capacity of the unit plus 10 per cent. Cable connections should be monitored for phase correctness and, before energising, a check carried out that test certificates (including earthing) are complete and available; that any breather openings have been uncapped and any breather dryers (if fitted) are in good condition; and that ventilation is adequate.

Where work to old equipment is involved, the Clerk of Works should verify that the transformers do not contain any hazardous materials. Legislation now forbids the use of polychlorinated biphenyls (PCBs), and this highly toxic and environmentally destructive chemical must be withdrawn from service. If such a hazard is discovered, all work in the vicinity should be prevented and the A/EA and project engineer informed.

7.17.7 Switchgear

Owing to the wide variety of HV switchgear available and the need to select carefully to suit requirements, any changes proposed from the specification must be referred to the project engineer; this applies as much to the protection equipment as to the switchgear itself. The manufacturer's detail schedule should be verified. If the switchgear is to operate in conjunction with any existing switchgear, for example cable inter-tripping, confirmation should be obtained on whether protection devices are compatible, although this element should be done by the project engineer.

All switchgear must be installed with sufficient room in front for removal and maintenance of component items (circuit breakers with trucks may require 2,000mm); alternatively, positioning switchgear opposite a door opening of sufficient width may be acceptable.

There are a number of specifications pertinent to electrical switchgear depending on the operating voltages and locations. Advice should be sought from a specialist in this field if there is any doubt whatsoever regarding the suitability of switchgear for the intended job or location.

7.17.8 Standby generators

For installation, testing and commissioning purposes, HV generators generally follow the same procedure as LV sets, with the exception that the increased voltages require additional safety requirements and a fully detailed method statement signed by an AP and countersigned by a second AP. All procedures, and particularly those regarding 'high voltage enclosures', must be strictly observed.

If the generating set is to be built up on site, the Clerk of Works should liaise with building and civil engineering (B&CE) colleagues to verify that any concrete foundations are in accordance with the specification before the erection team arrives on site. Additional rotational lineability checks will be needed before running the set to ensure that the prime mover and generator are in line and will not put undue stress on the coupling or create vibration.

The Clerk of Works must check for adequate ventilation and that attenuation is fitted in accordance with the project specification.

7.17.9 Substations

The Clerk of Works should verify the builders' work requirements and check that adequate allowance has been made to accommodate the size and number of cables to be installed. In particular, the turning radius of cables should be given adequate consideration, especially at and between switchgear. Bases and plinths should be designed to accommodate cables, incorporating cut-outs or ducts as required. Consideration must be given to the containment of oil that may escape from oil-filled transformers in the event of accidental spillage. Where substations are to be surrounded by walls or are in buildings, adequate provision must be made for the escape of personnel in the event of fire or other emergency.

The Clerk of Works must verify that adequate allowance has been made in the planning of the substation for the removal and maintenance of equipment, access for cable jointing, withdrawal of circuit breakers and any future reasonable extension of switchgear. In case of any doubts, the A/EA and health and safety coordinator should be consulted.

Particular attention should be paid to the requirements for earthing of the substation equipment, and the contractor must be made aware of all obligations under the contract.

7.17.10 Testing

All tests required by the specification and the project engineer must be witnessed; all salient dates should be recorded in the site diary and a copy of all test results retained.

All HV pressure testing should be witnessed; all test equipment must be correctly calibrated and accompanied by current certificates of calibration. Test areas should be defined by temporary barriers and warning notices posted. Test voltages and duration should be in the job specification or as advised by the project engineer in writing. For jointing onto existing cables or equipment, test voltages and/or duration are reduced; further details are provided in the appropriate British Standards for the type of cable.

7.17.11 Identification labels

All identification labels for switches, circuits, cable boxes, etc. must be of a permanently engraved type with lettering at least 10mm high and mechanically fixed, i.e. not using adhesives.

7.17.12 Safety and identification signage

Adequate and sufficient identification and danger notices must be fitted to equipment and substations, and must be visible from the normal direction of access.

Quality control: low voltage – distribution, lighting and power

7.17.13 Definitions

Under the IET Wiring Regulations, BS 7671, low voltage (LV) is defined as:

- not exceeding 1,000V AC or 1,500V DC between conductors
- 600V AC or 900V DC from any conductor to earth.

7.17.14 IET Wiring Regulations

All requirements for low and extra-low voltage power design installation and control are contained in the IET Wiring Regulations BS 7671:2018 – and are enforceable by law in England and Wales through the Electricity at Work Regulations 1989 for commercial premises and through the Building Regulations in Scotland. The regulations are a joint publication between the IET and the BSI.

The IET publishes the regulations in paperback and online digital formats and also publishes guidance notes for design, installation and testing. The Clerk of Works should request copies of all relevant publications from the A/EA or project engineer.

7.17.15 Design

Although the designer may have detailed the cable sizes and circuit protection devices, the electrical contractor must produce working drawings showing the cable routes, builders' work and all other items needed to install the circuits. This should include locations where fire-stopping is required. The electrical contractor should also check the calculations in case the actual route or installation method chosen produces a value requiring a different size of conductor or protective device. These drawings should be submitted for approval prior to construction. Consideration should be given to energy efficiency in the design and sizing of cables.

7.17.16 Revised voltage ratings in Europe

Because of the adoption of revised voltages across Europe, at 230V for single phase, 400V for three phase, it is now possible to install equipment, luminaires or lamps working at the reduced voltage on existing 240V/415V circuits. The Clerk of Works should be aware of the problems that can be created by long-term use of lower-rated equipment and endeavour to ensure that 230V and 240V goods are not inter-mixed. If 230V/400V goods are installed throughout the job the Clerk of Works should endeavour to ensure that the supply transformer is adjusted to produce the correct output by tap changing. Any problems should be referred to the project engineer. Although the voltage still falls within UK tolerances, it is advisable to check that distribution boards and operating/maintenance manuals are annotated to show that the system has been designed to operate at 230V/400V, and any lamps and spares used or alterations to the system must be for the correct voltage. Any problems should be raised with the A/EA and the project engineer.

7.17.17 Buried services

Any buried services must be handled so as to avoid damage and in accordance with manufacturers' recommendations; also, the correct preparation and backfilling of trenches, together with specified warning tapes, tiles, etc. must be carried out. The covering and segregation of cables and other services must be adequate. In particular, compliance with the cable manufacturer's instructions as to the laying temperatures and minimum bending radii must be monitored. The exact positions of the cables and any joints should be recorded for later checking of the record drawings.

7.17.18 Buried conduits

The Clerk of Works should inspect conduits pre-installed for embedding in slabs or screeds before they are hidden, checking for site damage and corrosion, protection of cut threads and frequency of conduit boxes. Checks should also be made for electrical continuity where the conduit is to be used as a protective conductor (although this is not currently the usual method). The cover to buried conduits must be adequate, and fixings must be secure to prevent movement during pouring and vibration of concrete. Particular care should be taken where conduits cross expansion and settlement joints, and the contractor should be consulted on this matter. Open ends of conduits should be temporarily plugged immediately to prevent the ingress of water and solid materials. Before cables are drawn in, conduits must be cleaned internally.

During first fix the Clerk of Works should verify that chases are not cut too deep. Building Research Establishment (BRE) guidance notes recommend that they are not greater than one-third of the thickness of the block or brick, but are sufficient to allow the correct amount of plaster cover. Suitable fixings for the wall construction material must be used. The Ministry of Defence (MOD) *Specification 034 Electrical Installations* has details for work on Crown Property. The project engineer or architect should be consulted for other types of contract.

7.17.19 Cable supports

All cable supports should be inspected for adherence to the specification, ensuring that fixing centres are adequate. In particular the cable manufacturer's recommendations with regard to the support of vertical cables should be noted. Particular attention should be paid to ensuring that the contractor allows for the bending radii of cables when planning tray work or cable rack routes. Cut edges of cable trays and cable rack are to be deburred and given a coat of zinc-rich paint. Sharp edges should be fitted with protection for the cables. Coupling bolts should not come into contact with the cables, and all cables should be fastened with proprietary ties. Attention should be drawn to the requirements for supplementary bonding. Cable ties for fire alarm cables must be non-flammable (e.g. stainless steel).

7.17.20 Cables

Cable drums must be unloaded carefully and properly stored in accordance with the manufacturer's recommendations. Cables must be handled with care, and every effort made to avoid damage to them, to other services and to the building fabric. Particular attention should be paid to the manufacturer's recommendations with regard to bending radii, and the way in which cables are pulled into conduits, ducts, etc., should be monitored. Any damage to cables should be noted in the site diary and reported to the contractor and A/EA. The size of conductors and type of cable should be checked against the design drawings.

7.17.21 Installation of cables within the fabric of buildings

Where cables run under a floor or in a ceiling, cables passing through or on joists or battens must do so at least 50mm vertically from the top and/or bottom as appropriate. Where cables are installed in walls and/or partitions at a depth less than 50mm, cables are to be installed within 'safe zones'. Zones are within 150mm from the top of the walls and/or partition or within 150mm of an angle formed by two adjoining walls and/or partitions, and also horizontally and vertically of any accessory. Irrespective of the depth of concealed cables, where the construction is made up of metallic cables it must be protected by a residual current device (RCD). Refer to the IET Wiring Regulations 522.6.

7.17.22 Thermal insulation

The Clerk of Works should ensure that the derating factor of thermal insulation exceeding 100mm is taken in to consideration. Particular attention should be given in domestic installations to electrical showers, ovens and ring final circuits.

7.17.23 Conduits, trunking and ducting

Unless routes are indicated on drawings, their run should be determined with the contractor before work is started, and the routes agreed with the project engineer on behalf of the A/EA. The Clerk of Works must verify that the contractor pays due regard to the pulling in or laying of cables and subsequent access for future maintenance, paying particular attention to the segregation between other services and cable systems. Conduit, trunking and ducting runs should be run parallel with the lines of the building construction. Any sharp edges and burrs are to be removed as the works progress, a zinc-rich paint applied to all cut edges, and adequate attention paid to allowing free passage for cables.

The Clerk of Works should check on trunking and ducting that earthing links are present. Refer to the IET Wiring Regulations for cable-carrying capacities of conduits, trunking and ducting. Before any wiring is carried out, all conduits, trunking and ducting runs must be inspected and tested in accordance with the specification. Adequate capacities with regard to category segregation should be confirmed by reference to the Regulations.

IET Guidance Note 1 and *Onsite Guide* contain a number of useful tables and diagrams relating to the topics covered from 7.17.17 to 7.17.23.

7.17.24 Wiring accessories

The Clerk of Works should ascertain that wiring accessories have the required minimum degree of protection, conform with the specification and are mounted at the correct heights.

7.17.25 Ring final circuits

As a general rule ring final circuits must have a minimum conductor size of 2.5mm^2 on PVC cables and 1.5mm^2 for mineral insulated copper cables. Where there is thermal insulation or grouping of cables it will be necessary to ensure the minimum current carrying capacity of the conductors is not less than 20A.

7.17.26 Connections to appliances and motors

The Clerk of Works should pay attention to vibrating and rotating plant, machinery or motors and verify that adequate allowance is made for movement and mechanical stresses. Equipment intended to be moved must be connected by flexible cables unless supplied by contact rails. Flexible cables must also be used to supply stationary equipment where such equipment can be moved temporarily for cleaning or disconnecting. Consideration should also be given to maintenance access in fixed equipment such as boilers, particularly where terminations are made on moving access panels or doors.

7.17.27 Luminaires and lamps

The situation with regard to the lamping of luminaires should be clarified prior to handover. It should be established whether the specification or contract allows for the contractor's use of the luminaires as temporary lighting, or if the A/EA is required to give permission. The Clerk of Works should verify the implications for relamping prior to handover have been agreed and understood.

Adequate supports must be available for suspended luminaires. In particular, the requirements where luminaires are to be incorporated into a suspended ceiling must be noted.

7.17.28 Emergency lighting

Refer to BS 5266-1, BS EN 1838, the project specification and MOD *Specification 034* (if applicable) and any building control officer's requirements for escape lighting, battery cubicles and testing. The provision of mains failure simulation must be monitored.

7.17.29 Identification labels

All labels and distribution schedules must be fitted in accordance with IET Wiring Regulations, and the Clerk of Works must verify these reflect the 'as installed' systems, including any construction stage alterations.

7.17.30 Generators

Low voltage generators are normally delivered as skid-mounted package units that have been assembled and tested at the supplier's works. Site tests should be conducted as required in the specification and against the specified test load, i.e. a load bank.

If the generator is of a new type or from a new supplier, tests at the manufacturer's work or under controlled environmental conditions may be required. The project engineer should decide whether the Clerk of Works will be called on to witness trials on their behalf. The A/EA should be consulted if the location for tests is any distance from site, as there may be others better placed to visit the test site.

All fire and safety devices must be installed and working correctly before the set is run. Section 7.19 below has further information regarding fuel systems.

Adequate air requirements and noise attenuation must be installed to specification.

7.17.31 Testing

All testing and inspection is to be in accordance with IET Wiring Regulations. All instruments must be in good condition and accompanied by certificates of calibration, even if newly purchased. Forms for results are to be based on the model forms in the IET Wiring Regulations. The Clerk of Works should ensure witness testing takes place to confirm that correct inspection and testing procedures are followed. Results are to be recorded to the Electrical Installation Certificate or checked against those already recorded.

Quality control: specialist supplies and distribution

7.17.32 Computers: 'off standard'

Some computer systems may be powered from sources that contain back-up sources or special electrical filters to eliminate surges, interference, etc. Frequently, plug-in power points on such circuits are 'off standard' to prevent other appliances being plugged in. The socket outlets and plugs should comply with BS 1363 in all respects except that the shape and/or orientation of one or more pins of the plugs and corresponding socket tubes should be different from those on standard plugs and sockets.

Each socket should be labelled to indicate that it is a dedicated supply intended only for computers or similar equipment. See section 7.17.33 below for information on 'clean earths'.

7.17.33 Filters

Installations and equipment that require a 'clean' power supply free from electrical interference, etc. may be fed from an installation containing filters of various types. These remove interference, harmonics, spikes, etc. within defined ranges and tolerances. If required to be installed, the specified model should be strictly adhered to, as variation could result in a failure to meet the requirements. The project engineer should be consulted in case of any doubt.

All earths from such filters should be run independently of any building or electrical earths to prevent cross-interference.

7.17.34 'No break' supplies

'No break' supplies and uninterruptible power supplies produce a similar result by various mechanical and electrical processes. Uninterruptible units are normally semi-sealed units converting the AC power to DC and then back to AC electronically, with a battery source connected at the DC point in the system. These can range from a small under-table unit for a PC, up to modular units capable of supporting considerable loads (in excess of 1MW). The battery back-up is normally of the sealed or low-maintenance type and capable of supporting the uninterrupted power supply (UPS) at full load for 15 minutes. A bypass circuit supplies 'mains' for use in case of breakdown.

Mechanical interface uninterruptible units are often specified where electrical separation is desired by the client between the client's equipment and its electrical power source. They can consist of a motor and generator on a common shaft with a heavy flywheel between to maintain momentum until a standby generator on the motor supply is able to take the load. Alternatively, a solid-state rectifier supplying DC to a motor connected to a generator can be used. In this case, a battery source maintains the DC supply until the standby generator is online.

Depending on the specific site requirement, either multiple units may be provided to allow for maintenance or a bypass switch provided.

All performance tests should be carried out at the specified load rating.

All test instruments should be accompanied by current statements of calibration. Dates should be recorded in the site diary and a copy of all test results retained for site records.

There are many variations on the very basic outlines described above. All installation testing and commissioning should be carried out in accordance with the equipment manufacturers' recommendations and method statements.

The Clerk of Works should be aware during testing that the DC voltages can be higher than usual voltages encountered; also, the large number of capacitors fitted in UPS units are an additional hazard. Sufficient time must be allowed for the charge to dissipate, or it may be discharged using an earthing stick. The Health and Safety Plan and File at handover must contain safety statements/ procedures.

7.17.35 Uninterrupted power supply

Any special requirements of the project specification with regard to earthing, ventilation and cooling of the UPS and battery rooms should be noted. Battery racks must be stable and properly secured. Generally, racks should be of acid-resistant material, which is likely to be plastic-coated steel. Particular attention must be paid to protection of the coating.

There may well be specific requirements with regard to off-site testing and, in the case of larger units, extended on-site testing. If off-site testing is witnessed by others, the Clerk of Works must verify test certificates that accompany the equipment. The requirements for witnessing should be confirmed with the A/EA, and the Clerk of Works should confirm that test equipment, artificial electric loads and any other necessary facilities will be available at the required time.

7.17.36 Special earthing

For some installations the client may specify a 'clean earth'. This is run as a separate insulated earth (protective conductor) from the connection point of the equipment or earth pin of a dedicated 'off standard' socket outlet, to a point as near as possible to the source of electrical supply (e.g. the main earth bar). All other electrical earths are as standard.

Occasionally, for UPS and power filtration or conditioning equipment, a separate insulated 'earth' stud will be fitted. This is a 'dump point' for the unwanted electricity components (e.g. spikes, harmonics and surges removed from the supply) to produce a clean sine wave. This should be treated in the same way as a 'clean earth'. Sometimes the voltage produced from this point can amount to several volts and may need to be routed to a separate earth rod located outside the main earth so that their zones of resistivity do not overlap.

If problems are encountered with sensitive electronic equipment malfunctioning, the earth system should be tested and monitored as the presence of voltages as low as 5V can cause problems.

7.17.37 Extra-low voltage (ELV) supply

The IET Wiring Regulations make a subtle difference between safety extra-low voltage (SELV) systems and functional extra-low voltage (FELV) systems, which relate to the protective measures taken. The Clerk of Works should ascertain which applies to the systems in use. The change to the definition of nominal voltage should also be noted. ELV circuits must be segregated from other types of circuit in accordance with the requirements of the IET Wiring Regulations.

7.17.38 Builders' work

Builders' work holes, ducts, plinths, etc. for M&E items must be verified to ensure that these are adequate for their duty, correctly located and aligned.

Quality control: specialist lighting

7.17.39 Application

This section covers all lighting other than general lighting and emergency lighting, which are dealt with in the section above.

7.17.40 Design

Specialist lighting is normally designed, either by the equipment manufacturer or a person who specialises in such work, to the specific light output in terms of efficacy, colour rendering and light spread for a specific luminaire and lamp. Any alterations proposed by the contractor must be referred to the designers for their consideration. It should be borne in mind that even a change of lamp can create a change in appearance if the colour rendering is different. A change in manufacturer can sometimes cause unanticipated problems in this respect.

7.17.41 Low voltage display lighting

LV display lighting requires close monitoring during installation, as the higher currents required by the lower voltages may need wiring to be of a larger conductor size and/or higher temperature cables. Additionally, if the LV luminaries are installed on lighting trunking, the trunking power rating will need to be increased considerably.

For example, while the voltage of a 12V lamp is 20 times less than normal, the current rating will need to be increased by 20 times for the same power, so a trunking feeding six 50W 12V lamps will need to be rated at 25A!

There are special requirements for neon signs. Where the operating voltage of the sign exceeds 'low voltage', an emergency switch must be provided. BS 7671:2008+A3:2015 provides further details.

7.17.42 Luminaires generally

The Clerk of Works should verify that luminaires comply with the specification, paying particular attention to the colour of lamps. The luminaires must be positioned in accordance with the contract and working drawings, they must coordinate, and they are not to be obscured by other services. The contractor should be encouraged to provide a coordinated ceiling plan indicating luminaires and diffusers, etc. to resolve any problems at the outset. If a clash with other services is noted, this should be brought promptly to the attention of the A/EA and M&E project engineer.

7.17.43 Aviation ground lighting

Aviation ground lighting (AGL) is a singularly different facility to most other lighting installations. Because of the need to maintain constant levels of brilliance throughout a circuit, it is frequently arranged as a series circuit with each luminaire being connected to the primary circuit through an isolating transformer. This arrangement means that a relatively low current is used (e.g. 6.6 or 8.33A) but the number of luminaires used can result in a voltage considerably higher than 1,000V.

For this reason AGL control centres are restricted areas, and should only be entered by people authorised or accompanied by the authorised person. All cabling used on the luminaire side of the control transformers must be 2,000V and rated and colour coded for the length of its outer sheath, using the colour specified in the design or maintenance manual. All earthing conductors should be bare stranded copper of 6mm minimum size and buried directly in the ground. The Clerk of Works should verify that cabling installed in trenches is not laid as a complete ring to control centres to avoid inducing currents via 'transformer effect' into other circuits.

As the circuits run from the control centre to the nearest luminaire and then interlace, no cable joints should be allowed between luminaires, and only if essential elsewhere. Sufficient slack should be allowed on circuits at each isolating transformer to enable its removal clear of ground level, plus at least 300mm.

Warning: for alterations to an existing system on circuit, the authorised person should identify, cut and prove the circuit to be worked on, securely isolate the circuit and issue a permit to work. Colour identification must not be relied on, as the use of incorrectly coloured cables for repairs is all too common.

The Clerk of Works should not permit work on an existing airfield lighting circuit unless the electrical contractor's competent person is in possession of a permit to work. Details of the issue and cancellation of any permits should be recorded in the site diary. Any interruptions to work caused by activities beyond the contractor's control, for example flying, military activities, etc., should also be recorded, as should all testing and test results.

The design and spacing of AGL is controlled by the Civil Aviation Authority for civil airports, and for military airfields it is derived from STANAG 3316; both of these requirements comply with internationally agreed standards. The Clerk of Works must not agree any alterations proposed by the contractor unless authorised by the project AGL design engineer and client. The spacing of luminaires must be confirmed to be correct and, where appropriate, it must be confirmed that the setting out dimension is derived correctly.

Some items of an AGL installation, for example precision approach path installation (PAPI) and inertial referenced flight inspection system (IRFIS), are positioned precisely and are used by aircraft in conjunction with other navigation aids. The setting out and positioning of such equipment is best done by experienced surveyors.

Generally, further installation information on AGL can be found in MOD *Specification 034*. If any difficulty arises or if advice on AGL installations is required, the A/EA should be requested to organise specialist help.

7.17.44   The industrial environment

For further reading, refer to the *CIBSE Lighting Guide LG1*.

Luminaires in hazardous areas should be checked to ensure that the correct ingress protection (IP) rating and electrical classification have been applied (e.g. 'Buxton' rating for flameproof fittings). The Clerk of Works should ensure that any flameproof requirements for luminaires apply to associated wiring installations.

Where high bay luminaires are specified, due consideration must be given to relamping and maintenance. The Clerk of Works should verify that the Health and Safety Plan and Health and Safety File at handover contain safe procedures.

7.17.45 Hospitals and healthcare buildings

For further reading, refer to the *CIBSE Lighting Guide LG2*.

Luminaires in hazardous areas should be checked to ensure that the correct IP rating and electrical classification have been applied.

Many areas will have lighting controlled by dimmer switches, and the Clerk of Works should ensure that these are correctly selected and operate satisfactorily. Operating luminaires may incorporate pattern selection, and a check should be made that this function operates correctly and that the luminaire focuses as required. Colour rendering may be particularly important in some areas. Special switching arrangements will apply in radiological areas, X-ray darkrooms, etc.

7.17.46 Areas for visual display terminals

For further reading, refer to the *CIBSE Lighting Guide LG3*.

Lighting will have been designed to eliminate high luminance from the room or to ensure that any high luminances that are present are not reflected towards the operator. This may be achieved by downlighters, fixed or freestanding uplighters, or a combination of these.

7.17.47 Sports lighting

For further reading, refer to the *CIBSE Lighting Guide LG4*.

Luminaires may be difficult to reach, especially with mounting heights of up to 10m. The Clerk of Works should verify that adequate consideration has been given to the maintenance and relamping of luminaires with due regard to the Health and Safety at Work etc. Act 1974 and CDM 2015. The Health and Safety Plan and Health and Safety File at handover must contain safe procedures.

7.17.48 The visual environment in lecture, teaching and conference rooms

For further reading, refer to the *CIBSE Lighting Guide LG5*.

Lighting schemes may incorporate luminaires for general display and presentation lighting, and may be under the control of sophisticated lighting consoles. Because there may be a greater than normal number of luminaires and other services (e.g. ventilation ductwork fitted in suspended ceilings), coordination of services is of the utmost importance.

Because of their location in large auditoria and over-raked seating, luminaires may be difficult to reach. The Clerk of Works must verify that adequate consideration has been given to the maintenance and relamping of luminaires with due regard to the Health and Safety at Work etc. Act 1974 and CDM 2015. The Health and Safety Plan and Health and Safety File at handover must contain safe procedures.

7.17.49 The outdoor environment

For further reading, refer to the *CIBSE Lighting Guide LG6*.

Lighting will not only be provided for the illumination of walkways and roads, but also for the display of architectural features or sculptures. Particular attention should be paid to the siting of luminaires, to the IP and electrical classification of fittings, and to electrical safety.

Depending on their mounting heights and locations, luminaires may be difficult to reach. The Clerk of Works should verify that adequate consideration has been given to the maintenance and relamping of luminaires with due regard to the Health and Safety at Work etc. Act 1974 and CDM 2015. The Health and Safety Plan and Health and Safety File at handover must contain safe procedures.

Quality control: electric heating

7.17.50 General

All appliances must be examined to verify that they comply with the IET Regulations. In particular, attention must be paid to appliances that may be installed in hazardous areas. The Clerk of Works should verify that individual heaters can be isolated for safety and maintenance, and that cables are suitably heat resistant where required.

7.17.51 Radial circuits

As no diversity is permitted for heating installations permanently connected into an electrical circuit, heating appliances that are heavily used should be wired independently as radial circuits to avoid overloading circuits under normal use. (This applies for example to cookers, all types of water heater, hand dryers, storage heaters and all types of space heating not designed for occasional use only.)

7.17.52 Embedded heating

Under-floor heating installations use a higher resistance cable than standard cabling, usually with a double-sheath single-core unarmoured cable. Care should be taken during installation to ensure that the outer sheath is undamaged and is located correctly, and that later operations do not cause damage. Reference to electrical under-floor heating systems can be found in the special locations section 735 of the IET Wiring Regulations.

Note: where normal electrical circuits pass through a heated floor, they must have the appropriate ambient temperature correction factor applied.

7.17.53 Water heaters: with immersed or uninsulated elements

Water pipes supplying water heaters are to be solidly and metallically connected to the water heater. The water pipe must also be connected to the main earthing terminal via an independent circuit protective conductor. Power to water heaters must be directly connected via a double-pole isolator and cannot be connected using any form of plug and socket arrangement. Refer to 554.3 of the IET Wiring Regulations

7.17.54 Instantaneous water heaters

Refer to the IET Wiring Regulations section 554.1.

Instantaneous water heaters are unsuitable for use on installations where a water softener of the salt regenerative type is used, because the increased conductivity of the water is likely to lead to excessive earth leakage currents from the usually uninsulated element. It is essential that all parts of this type of water heater are solidly connected to the metal water supply pipe, which in turn is solidly earthed independently of the circuit protective conductor. If the neutral supply to a heater with an uninsulated element is lost, current from the phase will return via the water and the earthed metal. Therefore, a careful check is necessary to ensure that there is no fuse, circuit breaker or non-linked switch in the neutral conductor.

7.17.55 Water heaters: non-vented

The use of unvented water heaters for multi-point purposes requires special connections to ensure that a safe working pressure is not exceeded. Refer to BS EN 60335.

7.17.56 Electrode boilers and water heaters

Two or three electrodes are immersed in the water, and a single-phase or three-phase supply is connected to them. There is no element and the water is heated directly by the current flowing between the electrodes. The electrical requirements are as follows.

- **For three-phase:** a circuit breaker that opens all three phases and is equipped with overloads on all phases; bonding conductors equal to the size of the supply conductors (or larger) and connecting the heater shell to the armouring on sheath of the supply cable and the heater shell to the neutral conductor.
- **For single-phase:** double-pole linked circuit breakers with overload protection on phase and neutral. Bonding conductors to be at least the size of the conductors between the shell of the heater to the sheath, armour or earth of the supply cable, and the shell of the heater to the neutral. The supply must have an earthed neutral.

- **For three-phase heaters fed from extra-low voltages:** special conditions apply. Refer to BS 7671 for information.

7.17.57   Cookers

The cooker control unit should be selected in accordance with the IET Wiring Regulations section 553.

7.17.58   Locations containing a bath or shower

Means of isolation for heaters in locations containing a bath or shower, for example instantaneous shower heaters, must be sited out of reach except by pull cord, and should be of specific type. Shaver sockets must be fed from a double-wound transformer to BS EN 61558-2-5, and protected by an RCD with an operating current no greater than 30mA. Extract fans should have a three-pole isolator located outside the area. The IP rating of equipment must be selected in accordance with the zone in which it is installed. (Zone 0: IPX7, Zone 1 and 2: IPX4.)

7.17.59   Heater batteries

The requirements of the contract documents should be verified with regard to the switching, control and construction of heater batteries. All wiring must comply with the IET Wiring Regulations and the Clerk of Works should verify that any overcurrent protection required by the contract documents are provided.

Statutory requirements and technical standards

Publications to which reference may be necessary include:

Electricity at Work Regulations 1989. SI 1989/635
Electricity Supply Regulations 1988. SI 1988/1057. MOD Specification 034 Electrical Installations
Health and Safety at Work etc. Act 1974. (Chapter 37)

HSE publications

ESI (Electrical Supply Industry) Standards. ESI 09-9 Waveconal Cable
GS 6 Avoidance of Danger from Overhead Electrical Lines
GS 38 Electrical Test Equipment for use by Electrician
IEC (International Electrical Commission) 502 – XLPE cables

Institute of Electrical Engineers publications

Guidance Notes 1 to 8 to the IET Wiring Regulations BS 7671
On Site Guide to the IET Wiring Regulations BS 7671

British Standards

BS 1363 13 *A Plugs, Socket-Outlets and Adaptors (Various Parts and Dates)*
BS 5467:2016 *Electric Cables. Thermosetting Insulated, Armoured Cables of Rated Voltages of 600/1 000V and 1 900/3 300V for Fixed Installations. Specification*
BS EN 60076 *Power Transformers (Parts 1 to 20, Various Dates)*
BS EN 61558-2-5:2010 *Safety of Transformers, Reactors, Power Supply Units and Combinations Thereof. Particular Requirements and Tests for Transformer for Shavers, Power Supply Units for Shavers and Shaver Supply Units*

Electrical switchgear

BS 7657:2010 *Specification for Cut-out Assemblies up to 100A Rating, for Power Supply to Buildings*
BS EN 60947-2:2006+A2:2013 *Low-Voltage Switchgear and Controlgear. Circuit-Breakers*
BS EN 60947-3:2009+A2:2015 *Low-Voltage Switchgear and Controlgear. Switches, Disconnectors, Switch-Disconnectors and Fuse-Combination Units*
BS EN 60947-4-1:2010+A1:2012 *Low-Voltage Switchgear and Controlgear. Contactors and Motor-Starters. Electromechanical Contactors and Motor-Starters*
BS EN 60947-4-2:2012 *Low-Voltage Switchgear and Controlgear. Contactors and Motor-Starters. AC Semiconductor Motor Controllers and Starters*
BS EN 60947-4-3:2014 *Low-Voltage Switchgear and Controlgear. Contactors and Motor-Starters. AC Semiconductor Controllers and Contactors for Non-Motor Loads*
BS EN 60947-5-1:2004+A1:2009 *Low-Voltage Switchgear and Controlgear. Control/Circuit Devices and Switching Elements. Electromechanical Control Circuit Devices*
BS EN 61439-2:2011 *Low-Voltage Switchgear and Controlgear Assemblies. Power Switchgear and Controlgear Assemblies*
BS EN 61439-3:2012 *Low-Voltage Switchgear and Controlgear Assemblies. Distribution Boards Intended to be Operated by Ordinary Persons (DBO)*
BS EN 62271-1:2008+A1:2011 *High-Voltage Switchgear and Controlgear. Common Specifications*
BS EN 62271-100:2009 +A1:2012 *High-Voltage Switchgear and Controlgear. Alternating-Current Circuit-Breakers*
BS EN 62271-200:2012 *High-Voltage Switchgear and Controlgear. AC Metal-Enclosed Switchgear and Controlgear for Rated Voltages Above 1kV and up to and Including 52kV*

Low voltage

BS 5266-1:2016 *Emergency Lighting. Code of Practice for the Emergency Lighting of Premises*
BS 7671:2008+A3:2015 *Requirements for Electrical Installations (IET Wiring Regulations 17th Edition) (until January 2019)*
BS 7671:2018 *Requirements for Electrical Installations (IET Wiring Regulations 18th Edition)*

> BS EN 61439 *Specification for Low-Voltage Switchgear and Control Gear Assemblies (Various Parts and Dates)*
>
> Lighting
>
> As above, plus CIBSE Lighting Guides LG1 to LG12
>
> Electric heating
>
> BS EN 60335 *Specification for Safety of Household and Similar Electrical Appliances (Various Parts and Dates)*

7.18 COMMUNICATIONS, SECURITY, FIRE AND CONTROL SYSTEMS

Contract requirements

7.18.1 Duties

Conformity with specifications and drawings should be verified. The many variable performance requirements specified for products to be installed require particular vigilance in checking the documentation and certification of deliveries and conformity with appropriate standards, bye-laws and other statutory requirements. Competent personnel are obligatory for some tasks, and the Clerk of Works may be asked to confirm their qualifications.

The specification should be verified for applicable product QA schemes and the A/EA's instruction obtained on any duties arising.

The following sections contain contractual requirements specific to this subject.

Quality control: communications systems

7.18.2 Design and standards

Most communications systems are designed and installed by specialist companies to suit a performance specification and user requirements. Although individual components of equipment may be manufactured to various British Standards, the systems themselves are not covered in detail by any specific British Standards. BS 6701 is a code of practice for installation of apparatus intended for connection to certain telecommunication systems.

7.18.3 Government procurement

Some government departments (e.g. the Army and the RAF) have their own specifications for communications equipment and its enclosures. In fact, for many sites the contract is split between installation of a cable environment system and later installation of cables and devices by a specialist company, normally employed directly by the client after completion and handover of the construction project.

7.18.4 British Telecom, the Army and the RAF

British Telecom (BT), the Army and the RAF have their own specifications for the construction of external pit and duct systems, and further specifications for internal distribution.

For BT systems they also supply free issue for installation by the contractor of duct materials, draw-pit frames and covers, and building entry seals; the local BT external installation officer may be contacted via BT's general enquiries freephone number (refer to local directory).

Similar specifications and information should be obtained from alternative communications contractors and the A/EA or project engineer.

7.18.5 Cable installation

The Clerk of Works should ensure that the company contracted is the one to which the client has let the communications contract.

Within the building, communications cabling (Band I cables) must be installed separately from mains LV cables (Band II cables). This can be achieved by separate containment, physical barriers or compartmental cable ducts and trunking systems. Alternatively, the communication-cabling-conductors must be rated to the highest voltage present within the grouping of cables.

Underground cabling telecommunication must be segregated by a minimum distance of 100mm from underground power cables. Refer to the IET Wiring Regulations, section 528.

Where fibre-optic cabling is to be used, particular attention must be paid to bending radii and handling. Many types of fibre-optic cabling are fragile and must not be drawn through conduits, etc., which can lead to breaks in the cabling. Instead, most types of fibre-optic cabling must be carefully laid in.

All cable routes should have either cables or an approved draw cord (not for fibre-optics) left in them at handover.

7.18.6 Public address

The planning and installation of sound systems is detailed in BS 6259.

For installation on airfields and some other MOD establishments, there is a requirement for prioritised override operation of the system. The requirements should be detailed in the project specification. As prioritisation is achieved by terminal links, it is essential during commissioning that the Clerk of Works verifies that priorities are correct. There may also be sound tone generators for various warnings. The locations that can initiate the tones and areas to which they are broadcast will be in the specification or advised by the client.

If sound alarms are incorporated in other equipment, the Clerk of Works must establish that there is no conflict with site alarm tones: equipment alarms must be altered if necessary. The A/EA must be informed of any problems, and instructions sought.

If used for voice alarm, requirements may need to be covered by specification for Category 3 circuits, as defined by the IET Wiring Regulations.

7.18.7 Audio visual

Audio visual (AV) systems are advancing rapidly. New digital systems give a far higher picture quality and frequently use a mixture of fibre-optic and low voltage cable connection systems. It is difficult to give definitive guidance due to their rapid evolution, but the best point of guidance is the quality of the final sound or picture; any poor or loose connections will result in appreciable loss of quality. If the quality is not as good as expected from an incoming source, a portable signal generator should be connected as near as possible to the incoming point and used to check the internal system. See also Security systems, below (7.18.9 to 7.18.15).

7.18.8 Data

Data transmission within a building is normally done within a computer local area network (LAN) or similar system, with one control station and modem, or fixed line links between buildings and networks. The variations in style, type and quality of networks are considerable.

All such systems are designed and installed by specialists. Changes to the system once the original design has been agreed should be supported by confirmation that the design concept and method of operation is unaffected. Technology is evolving faster than specifications and designs. Data cables are classed as Band I cables, and the same considerations should be given to communication cables.

Apart from the need to ensure that the electrical supply is adequate, reliable and backed up by UPS, the design specification is the only recommended guidance document to the standards required. See also Building management systems, below (7.18.28 to 7.18.32).

Security systems

7.18.9 General

Building and property security can include a wide range of devices, including intruder detection systems (IDS), closed circuit television (CCTV) and access control/ entry systems.

The appropriate trade association is the National Security Inspectorate, which publishes codes of practice relevant to sections of the trade. The Department of Trade and Industry (DTI) also publishes standards related to some aspects of the industry, and some British Standards apply.

7.18.10 Specialist installers

For certain types of installation within some government buildings, a specialised security group must be employed. This is normally stipulated at design stage, and only the specified company must be used for reasons of departmental or even national security.

Usually, the electrical subcontractor will be expected to install the system containment (conduits, boxes, etc.) and the specialist will install and commission the sensing and activating devices at a later stage. Sometimes the system wiring will be by the specialist, or it can be part of the general electrical installation, but to the specialist's specification. In all cases, the Clerk of Works should verify the contract specifications and ensure that the demarcation between responsibilities is clear and unambiguous. Security systems generally will be Band I cable systems, and the same considerations should be given to communication cables. However, particular attention may need to be given to the voltage ratings of cables with the grouping.

Subject to the A/EA's agreement, it is suggested that a site visit by the specialist is arranged to take place at a suitable stage of the works to confirm positions of sensors, etc. before installation of conduits starts.

Segregation of electrical circuits should be undertaken in accordance with the IET Wiring Regulations: see also Communications systems, above (7.18.2 to 7.18.8).

7.18.11 Intruder detection systems

Cabling is likely to be run in secure conduits. Conduit runs should be coordinated with other services to avoid clashes. The Clerk of Works must agree the method of fitting detectors in doors and terminating cabling around them, and agree the location of motion sensors as instructed by the A/EA.

The coverage of any infrared detectors must be checked to ensure that they cover the required areas and that any blind spots do not compromise the ability of the system to detect intruders.

7.18.12 Closed circuit television

Cameras must be carefully positioned to ensure that only the client's property is in vision; coverage of parts of the public highway is acceptable, but not adjoining and adjacent private property.

The location and mounting positions of cameras should be agreed to give the best field of view, as instructed by the A/EA. Care should be taken with panning cameras to limit their travel, to avoid both obstructions and views of other property. A mechanical restrictor may need to be fitted. For pole-mounted cameras, it is important to bear in mind that access for maintenance must be practical (e.g. that access to water-wash reservoirs is possible without recourse to specialist access equipment, that folding columns do not obstruct access for emergency or other vehicles when lowered). The Health and Safety Plan should be consulted, and the Health and Safety File at handover must contain appropriate maintenance methods.

Home Office installations: special requirement apply; the particular specification must be consulted.

7.18.13 Entry systems

Card and cipher access, electronic locking and motorised doors are among the systems that may be encountered. The Clerk of Works should attend, if required by the A/EA, any off-site inspection and witnessing at the manufacturer's works, and agree the method of terminating cabling to electronic door locks, paying attention to the integrity of the security of the installation.

Access control systems that require the use of a keypad or insertion of a card should be positioned so that use by disabled people is feasible without undue effort or stress. Alternatively, an assistance call point may be acceptable if the work is an adaptation of an existing building. However, the Disability Discrimination Act 1995, which has been applied progressively since 1999, will mean that no person with a disability should encounter inferior conditions, facilities or services.

7.18.14 Training

Training should ideally take place at the manufacturer's premises in advance of the testing and commissioning on site. The Clerk of Works should liaise between the client department and the contractor as requested by the A/EA.

7.18.15 Testing and commissioning

The Clerk of Works should liaise with the client department responsible for the operation of the security systems, and invite their attendance during testing and commissioning.

Protection systems: fire and lightning protection

7.18.16 General

This section covers the installation of lightning protection, and fire alarm and detection systems.

7.18.17 Fire alarm cabling

Fire alarm cabling should be installed in accordance with the IET Wiring Regulations, having due regard for segregation of circuits: refer to 422.2.1, section 528 and section 560 of the IET Wiring Regulations. Where power is provided to fire alarms or similar safety services, cables must have a minimum one-hour rating in the absence of any other regulatory standard. In some commercial and industrial applications, mineral insulated copper cables (MICCs) will be used. Where the MICC has an outer covering, the Clerk of Works should verify that the copper sheath is protected by wrapping with plastic adhesive tape at terminations. Refer also to BS 5839 Part 1.

7.18.18 Lightning protection

For general provisions, refer to BS EN 62305, which covers all aspects of design, installation testing and maintenance.

During installation, the Clerk of Works should verify that all tapes are routed in such a manner that 're-entrant loops' are avoided (see paragraph 15.8 and figure 22 of BS EN 62305 Parts 1 to 4). Checks should be made that all fixings are correctly spaced (see table 1 of BS EN 62305 Parts 1 to 4), and it should be noted that the spacing varies according to the length of the run, and whether it is horizontal or vertical. If thermal fixing (e.g. 'cad welding' – a form of thermit welding) is to be used to join tapes, then all 'hot working' procedures in force on the site must be obeyed.

MOD *Specification 034* contains further details of materials and manufacture.

Where reinforcing bars are to be used as down conductors, or where other connection is specified to the building reinforcement, the Clerk of Works should liaise with the B&CE site inspector to ensure that this is incorporated during building construction. Tests of such connections are to be made before casting the structure, and test certificates retained for inclusion in the Health and Safety File. Where dissimilar metals are used, adequate precautions should be taken to ensure that corrosion does not occur.

Exposed down conductors should be routed to avoid sharp right-angle bends and overhangs.

7.18.19 Electromagnetic pulse protection

Electromagnetic pulse protection (EMPP) is only installed in rooms requiring this very specialised form of screening. Installation and proving can only be carried out by one or two companies.

The room is completely encased in a metal sheath with all components lap covered, and door openings also contain metal components within the seal to ensure total shielding. All incoming services must be routed through filters (for electrical circuits) or special glands (for non-electrical communications). Ventilation grilles and ductwork also have special design considerations. Within the enclosed area, all equipment and fixings are secured so that the integrity of the outer sheath is not impaired.

Testing is normally carried out by the installation contractor, but must be to the client's performance specification written for that particular project.

7.18.20 Testing and commissioning

Lightning protection systems should be tested in accordance with BS EN 62305.

Fire alarm testing and commissioning should be undertaken in accordance with the project specification. It is likely that the building control officer will need to be in attendance for the final testing and commissioning, but it is advisable that the M&E site inspector witnesses testing before that stage.

Control systems: general

7.18.21 General

Control systems for heating, hot water, air conditioning and ventilation can range from the simple to the highly sophisticated. MOD *Specification 036* contains outline information on the sensing points for various types of controller. The project specification will outline the type of control required, or a control philosophy and the manufacturer whose equipment was used as a basis for the design specification.

For gas-fired systems refer to 7.16, above, for technical requirements and standards.

Approved working drawings should be received on site in good time for them to be studied before installation starts.

7.18.22 Conduit trunking and trays

The Clerk of Works should ensure that routes are coordinated with other building services, paying particular attention to segregation.

7.18.23 Cable installation

Cable installation should be undertaken in accordance with MOD *Specification 034* and the IET Wiring Regulations. Particular attention should be paid to cable segregation and protection and handling if fibre-optic cabling is used (see 7.18.5, above). Flexible cables to detectors and motors, etc. should be kept as short as possible. Cable ends should be permanently identified.

7.18.24 Detectors

Particular attention should be paid to the location of detectors to check that they are placed in representative positions and in accordance with the manufacturer's instructions.

7.18.25   Control cubicles and panels

The Clerk of Works should ensure that all cubicles and panels are constructed to the degree of protection required by the contract documents and BS EN 60947-2:2006+A2:2013, and verify that all internal wiring is colour coded and bunched, and run on trays or within trunking. Cable ends should be permanently identified. Where the contract drawings are only indicative, the location of panels should be agreed with the project engineer and contractor. Off-site inspection may be required; the Clerk of Works should verify with the A/EA for attendance required.

7.18.26   Commissioning and performance testing

Before testing of the control system starts the Clerk of Works must ensure that all plant under its control is fully commissioned. With the A/EA's agreement, the people responsible for the maintenance of the facility after handover should be invited to attend.

During inspection and commissioning, it is essential that the proving of all operational and safety devices is witnessed. It should be noted that for some items, such as fusible links, actual operation is not practicable, but shorting out of electrical devices or operation of mechanical release buttons will prove the operational capability of the system.

High-temperature cut-outs must always be checked on domestic hot water services. Before commissioning, the Clerk of Works should verify the operating temperature required as indicated for domestic hot water, and endeavour to obtain a temperature outside the zone in which Legionella is considered to be a higher risk (i.e. outside the zone from 20 to 46°C). If it is higher than 60°C, the provision of warning notices (risk of scalding) should be considered at all outlets. Any doubts as to appropriate temperature should be referred to the project engineer.

Requirements for interfaces with other elements, for example lighting, fire/smoke alarms, security systems, etc. should be verified.

7.18.27   Training

The requirements of the contract documents for the training of client staff should be verified. Through the A/EA, the Clerk of Works should liaise as necessary with the client and contractor. It may be preferable for training to be undertaken at the manufacturer's works prior to commissioning and performance testing being carried out.

Building management systems

7.18.28   General

Building or energy management systems (BMS/EMS) may be stand-alone for each individual building, or may be networked to a central location or via modem to a remote location. They may incorporate a computer at the 'head end' where interrogation and adjustment of the system can be undertaken, and a local keypad control at each outstation, or require an interrogative computer to be plugged in at the outstation.

7.18.29 Technological change

BMS are designed, installed and commissioned by specialist companies. Developments and progress in computers and microchip technology mean that changes in equipment offered by manufacturers often render the job specification outdated before the project has reached the installation stage.

Conversely, new technology is often marketed before it has been fully tested and proved in practice, and to be offered the 'first installation of the new range' can bring with it totally unforeseen problems.

7.18.30 Cable installation

Cable installation should be undertaken in accordance with MOD *Specification 034* or the contract specification and the IET Wiring Regulations. Particular attention should be paid to cable segregation and protection and handling if fibre-optic cabling is used (see 7.18.5, above).

The Clerk of Works should liaise with the main contractor/mechanical subcontractors for supply connection to motorised valves, dampers and equipment.

7.18.31 Commissioning and performance testing

Before commencing testing of the BMS, the Clerk of Works must verify that all plant under its control is fully commissioned and monitor that sufficient personnel are in attendance to witness both the reactions at the head end and also at the plant being controlled. With the A/EA's agreement, the people responsible for the maintenance of the facility after handover should be invited to attend.

Often, the sheer size of a BMS installation and the number of sensor or control points will make it very difficult to carry out a 100 per cent check of the accuracy and effectiveness of them all. A 10 per cent sampling check spread across the project will normally be sufficient to provide a reasonable statistical check on the quality and accuracy. However, the actual sampling rate and acceptable pass/fail rate must be agreed with the project engineer and A/EA. Should the check produce unacceptable results, the A/EA and project engineer must be informed immediately and further instructions sought. It is suggested that either:

- the contractor is asked to recheck the system and provide evidence before the 10 per cent check is repeated, or
- the percentage sampling rate is increased.

The final check of working capability is for the system to run for a predetermined period (say 14 days) without a software or system failure. The system needs to be fully operational and all the equipment it is managing must be working normally for the reliability trial to be considered successful.

7.18.32 Training

The requirements of the contract documents for the training of client staff should be noted. Through the A/EA the Clerk of Works should liaise as necessary with the client and contractor. It may be preferable for training to be undertaken at the manufacturer's works prior to commissioning and performance testing being carried out.

Statutory requirements and technical standards

Publications to which reference may be necessary include:

General

BS 6259:2015 *Code of Practice for the Design, Planning, Installation, Testing and Maintenance of Sound Systems*
BS 6701:2016 *Telecommunications Equipment and Telecommunications Cabling. Specification for Installation, Operation and Maintenance*
BS 7671:2008+A3:2015 *Requirements for Electrical Installations: IET Wiring Regulations 17th Edition (until January 2019)*
BS 7671:2018 *Requirements for Electrical Installations: IET Wiring Regulations 18th Edition*
MOD Specification 034 Electrical Installations

Security systems

BS 4737 *Intruder Alarm Systems (Some Parts Continue to be Current, Various Dates)*
BS EN 50131-1:2006+A2:2017 *Alarm Systems, Intrusion and Hold-up Systems. System Requirements*
BS EN 60839 *Series Alarm and Electronic Security Systems (Various Parts)*
NACP20 *Code of Practice for Planning, Installation and Maintenance of Closed Circuit Television Systems*
NCP 30 *Code of Practice – Access Control Code of Practice for Planning, Installation and Maintenance*

Fire and lightning protection

As above, plus:

BS 4422:2005 *Fire. Vocabulary*
BS 5839 *Fire Detection and Alarm Systems for Buildings (Parts 1 to 9, Various Dates)*

BS EN 60730 *Specification for Automatic Electrical Controls for Household and Similar Use (Various Parts and Dates)*

BS EN 60947-1:2007 + A2:2014 *Specification for Low-Voltage Switchgear and Controlgear. General Rules*

BS EN 61558-1:2005 +A1:2009 *Safety of Power Transformers, Power Supply Units and Similar. General Requirements and Tests*

BS EN 61558-2.23:2010 *Safety of Power Transformers, Power Supply Units and Combinations Thereof. Particular Requirements and Tests for Transformers and Power Supply Units for Construction Sites*

BS EN 62305 *Parts 1 to 4 Code of Practice for Protection Against Lightning*

MOD Specification 036 *Heating, Hot and Cold Water, Steam and Gas Installations for Buildings*

Building management systems controls

There are British Standards that deal with the many elements used within any one BMS. Refer to the relevant standards depending on the range of elements used within any specific system, for example heating, electrical, air conditioning, refrigeration, etc.

7.19 TRANSPORT SYSTEMS: LIFTS, HOISTS, CRANES AND PETROLEUM, OIL AND LUBRICANT (POL) INSTALLATIONS

Contract requirements

7.19.1 Duties

Conformity with the specifications and drawings should be verified. The many variable performance requirements specified for products to be installed require particular vigilance in checking the documentation and certification of deliveries, and conformity with appropriate standards, bye-laws and other statutory requirements. Competent personnel are obligatory for some tasks, and the Clerk of Works may be asked to confirm their qualifications.

The specification should be verified for applicable product QA schemes and the A/EA's instruction obtained on any duties arising.

Sections below contain contractual requirements specific to this subject.

Installation of such equipment is normally carried out by specialists or manufacturers. The requirements of the contract documentation should be noted, in particular with reference to the attendance of specialists, competent people and insurers. The Clerk of Works should liaise with the A/EA in good time regarding when the presence of these specialists will be required.

Quality control: lifting equipment

7.19.2   Lifts

For lifts and other enclosed shaft machines, it is essential that the main structure is within tolerance before installation starts. Accuracy should be verified by the main contractor against the dimensions given in the specialist contractor's working drawings. If the shaft is outside specification, the A/EA should be informed immediately.

With the assistance of the B&CE site inspector, the preparation for the builders' work associated with the lift installation should be verified, with particular attention paid to verticality and alignment, landings, bases, plinths, channels and any cast-in items. A check should be made that no services installations or access routes, other than those provided for the lift equipment and personnel, share or pass through the machine room or lift well.

Provision must be made from the lift well for the ventilation, directly or by ducts, of smoke to the open air in the event of a fire. The lift motor room must be adequately ventilated. Provision must be made for the installation of a tubular heater to prevent frost attack of grease and oils.

Where the lift is to serve as a fireman's lift, the Clerk of Works should liaise, through the A/EA, with the building control officer, whose duty it is to inspect the operation of the installation.

Requirements in contract documents for automatic control for fire operation, for example all cars returning to the ground floor, should be checked, as should any intercom/alarm provision for occupants.

Maintenance tools must be provided and mounted by the contractor as required by the contract documents (in the motor room if possible).

7.19.3   Safety in lift shafts or other openings

During construction work and installation of the equipment, the Clerk of Works should verify that all openings have adequate physical barriers to prevent personnel or equipment falling into the shaft. During installation, this may mean a rigid fence or screen around openings to exclude all trades other than those directly involved in the installation. The Health and Safety Plan should be consulted for safe working procedures and the A/EA and the health and safety coordinator informed immediately of any concerns.

Adequate ladder access must be provided to the lift well. In the case of hydraulic lifts, the Clerk of Works should verify that a purpose-made prop has been provided to stop creep during maintenance and inspection, and that a suitably placed stop device has been provided in the well (see BS EN 81-20 and BS EN 81-50).

Attendance by a specialist insurance engineer will be required to witness load testing and testing of the safety devices. Via the A/EA, the Clerk of Works should liaise between the insurer's engineer and the contractor to arrange this attendance.

A check should be made to see that all fire-stopping requirements for services have been adhered to.

7.19.4 Safety in machine and pulley rooms

The Clerk of Works must ensure that contractors are aware of their responsibilities with regard to making provision for safe working in machine and pulley rooms in accordance with BS 7255:2012, particularly section 9.

7.19.5 Lifting beams

Where lifting beams are installed over lift shafts, or as part of a crane installation, these must be marked with safe working loads and any client register numbering required by the contract documents. Beams must be insurance inspected before handover, and this may require load testing. The Clerk of Works must ensure that the contractor is aware of their responsibilities and is in attendance when the specialist inspections take place.

7.19.6 Hoists

Document hoists will require the same attention as lifts for builders' work requirements, and attendance by the insurance engineer requires the same level of liaison.

7.19.7 Cranes

The Clerk of Works should verify builders' work drawings for any cast-in items, bases and plinths, and ensure that these are adequate and correct. They should also ensure that the contractor coordinates the installation of the other M&E services with the crane installation to avoid any clash of services. Testing and commissioning of the installation in accordance with BS EN 15011 should be witnessed in association with the insurance engineer and safety officer as necessary.

7.19.8 Test weights

The Clerk of Works must beware during commissioning and testing that floor loadings are not exceeded when test weights are brought in and out, and should check that the contractor is aware of any potential problems. Particular attention should be paid to potential damage to suspended access floors or other sensitive areas.

7.19.9   Statutory inspection by a competent person

All lifts and usually all lifting equipment are subject to statutory inspection by a competent person before being put into service, and must be inspected at regular intervals thereafter. Copies of all test certificates should be obtained and should be included in the Health and Safety File at handover.

Quality control: petroleum, oil and lubricant installations

7.19.10   Types of installation

Petroleum, oil and lubricant (POL) installations fall into several different categories: motor transport refuelling, aviation fuel storage and handling systems, vehicle lubrication and servicing, heating fuel storage, and fuel for standby generator sets.

Motor transport refuelling: generally for MOD and civil sites the main source of advice on design, construction and installation can be found in the Association for Petroleum and Explosives Administration (APEA) *Guidance for Design, Construction, Modification, Maintenance and Decommissioning of Petrol Filling stations*. There may be additional requirements to satisfy the local licensing authority regulations – these should have been taken into consideration at planning stage and should be contained in the particular specification.

Aviation fuel storage and handling: apart from a completely new installation, all work must be carried out following 'authorised person' (AP) and 'person in charge' (PIC) procedures, with the operating authority providing the AP and the contractor the PIC. Any equipment and pipe work, etc. to be worked on should normally be certified 'gas free' before work starts. The site inspector's role is limited to checking that the installation is as designed and to witnessing tests.

Due to the training, medical and material needs of such duties it is not envisaged that any site inspector will undertake AP or PIC roles. The Clerk of Works must make certain not to extend their duties or accept duties in such a way as to impinge on or interfere with AP or PIC duties.

Vehicle servicing bays and heating fuel installations: all necessary requirements are to comply with fire precautions. Control of pollution regulations and local authority requirements must be fulfilled, for example all bunds must be in place, pipe work identified and protected, tank alarms tested and operational, and fill points labelled with grade of fuel and capacity of tank.

7.19.11   General

When working in the proximity of existing installations the Clerk of Works should ensure that the contractor complies with the Health and Safety Plan and any permit to work system, and should be acquainted with the permit system and able to monitor the contractor's performance.

7.19.12 MOD authorised person

In MOD establishments an AP will have been appointed. At an early stage the Clerk of Works should make contact with the AP and continue to liaise with them throughout the work. When testing and commissioning pipelines, storage tanks, pumps and dispensing equipment, the attendance of the AP and of the client department that will take ultimate responsibility for the installation and equipment should be requested.

7.19.13 Storage tanks

Any off-site inspection and testing should be attended as required by the contract documents and the A/EA. If the off-site inspection is to be undertaken by others, the test certificate should be verified on receipt of the tank on site. The tank must be examined on delivery for damage to the shell, connections and any protective coating, in the latter case ensuring that remedial works are immediately undertaken to prevent any corrosion. Where the tank is to be buried, adequate holding down straps must be fitted and the tank filled prior to burying to prevent it 'floating'. The tank should not be emptied until the surrounding mass concrete has cured.

Where the tank is to be erected from sections or plates on site, then, in liaison with the B&CE site inspector, the Clerk of Works should verify that the builders' work base is correctly constructed and incorporates any holding down bolts and cut-outs.

7.19.14 Pipe work

The contract documents should be verified for the type and grade of material to be used in the construction of the pipeline. Any special requirements for welding or specific construction methods should be noted. Where welding is to be undertaken, the Clerk of Works should ensure that the operatives are competent (and hold current certification) in the particular welding procedures to be adopted. The levels and routes of pipe work must be checked to avoid coordination problems with following trades, and correct installation of any check valves, if required, must be ensured.

7.19.15 Fuel dispense columns

These may be client or contract supplied. Where the client is to provide the equipment, the Clerk of Works must confirm that it will be available when required by the contractor to avoid any delay. When commissioning the dispensing equipment, the client's representative and weights and measures representative should be in attendance, and the Clerk of Works should liaise as necessary through the A/EA.

Statutory requirements and technical standards

Publications to which reference may be required include:

Lifting installations

BS 2853:2011 *Specification for the Testing of Steel Overhead Runway Beams*
BS 7121 *Series (Various Parts and Dates) Code of Practice for Safe use of Cranes*
BS 7255:2012 *Code of Practice for Safe Working on Lifts*
BS 9999:2017 *Fire Safety in the Design, Management and use of Buildings. Code of Practice*
BS EN 15011:2011+A1:2014 *Cranes. Bridge and Gantry Cranes*
BS EN 81-20:2014 *Safety Rules for the Construction and Installation of Lifts. Lifts for the Transport of Persons and Goods. Passenger and Goods Passenger Lifts*
BS EN 81-50:2014 *Safety Rules for the Construction and Installation of Lifts. Examinations and Tests. Design Rules, Calculations, Examinations and Tests of Lift Components*
BS ISO 4409:2007 *Hydraulic Fluid Power. Positive-Displacement Pumps, Motors and Integral Transmissions. Methods of Testing and Presenting Basic Steady State Performance*
HSE *Guidance Note PM 55 Safe Working with Overhead Travelling Cranes*

POL installations

Association for Petroleum and Explosives Administration (APEA) Design, Construction, Modification, Maintenance and Decommissioning of Filling Stations, 2nd edition, 2005
BS 799 *Oil Burning Equipment (Parts 1 to 8, Various Dates)*
BS EN 13012:2012 *Petrol Filling Stations. Construction and Performance of Automatic Nozzles for use on Fuel Dispensers*
BS EN 13617-1:2012 *Petrol Filling Stations. Safety Requirements for Construction and Performance of Metering Pumps, Dispensers and Remote Pumping Units*
BS EN 13617-2:2012 *Petrol Filling Stations. Safety Requirements for Construction and Performance of Safe Breaks for use on Metering Pumps and Dispensers*
BS EN 14015:2004 *Specification for the Design and Manufacture of Site Built, Vertical, Cylindrical, Flat-Bottomed, above Ground, Welded, Steel Tanks for the Storage of Liquids at Ambient Temperature and Above*

7.20.0 CAVITY BARRIERS IN EXTERNAL WALLS

7.20.1 Rationale

At the time of this publication, the industry is still considering the full implications following on from the Grenfell Tower fire disaster. Consequently, whilst wholesale and far reaching change is anticipated, particularly in legislation, the use of materials and competency levels, it was believed prudent to consider specifically, the use of cavity barriers in external walls.

7.20.2 Fire Strategy

The issue of fire in buildings is both complex and multi-faceted. Consequently, the prudent Clerk of Works should ensure that they are fully aware of all the elements and interfaces that contribute to a fire safe building. i.e.:

- Cavity barriers and fire stops
- Compartmentalisation
- Fire stopping of penetrations
- Fire doors and door sets
- Fire safety risk assessments
- Active and passive measures
- Testing and verification requirements
- Specified materials, compliance and their correct installation

You should always refer to the requirements of the Building Regulations and associated guidance, including Manufacturers specific product literature and installation instructions.

7.20.3 Cavity Barriers and Fire Stops

Both items are elements of passive fire protection.

Fire Stop

- A passive fire protection element that is required to seal or close a discontinuity or imperfection of fit between building elements that are required to be fire resistant (e.g. compartment floors/walls) or is installed at any joint or junction in the fire protection element.
- This must provide the same degree of insulation and resistance to the passage of flame and smoke as the building elements into which it is installed. For example stopping that is installed between a slab edge and an external wall must have the same insulation properties and resistance to the passage of flame and smoke as the floor slab (e.g. 120 min integrity and 120 min insulation).

Cavity Barrier

- Barriers are used to close concealed spaces and prevent penetration of smoke or flame to restrict the movement of fire within a building. Cavity walls and ceiling voids would be considered concealed spaces and would generally be fitted with cavity barriers during construction.
- According to both BS 9991 and HYPERLINK "https://www.planningportal.co.uk/info/200135/approved_documents/63/part_b_-_fire_safety/2"   Approved

Document B2, cavity barriers are required to have 30 minimum integrity and 15 minimum insulation (BS 9991 Figure 24 and Table 3).

- Both documents also allow cavity barriers to provide only 30 min integrity; clause 19.2 of BS 9991 and cites 0.5mm thick steel and 38mm timber as being a suitable cavity barrier materials for the perimeter of windows (advice and substantiation should be sought from a fire engineer prior to any decision regarding omitting the 15 min fire insulation requirement from a cavity barrier element).

Both cavity barriers and fire stops must be correctly located and fixed in the method prescribed by the manufacturer. This includes ensuring the material proposed is appropriate for the space requiring to be sealed.

7.20.4 The Purpose of Cavity Barriers

A concealed space (cavity) in the external wall of a building can act as a chimney and provide an easy route for flame, hot gases and smoke to propagate from one compartment of a building to another. Unsealed cavities can allow air to be drawn in and smoke to vent out, enabling the fire spread to accelerate through the façade. This chimney effect enables flames that are within a cavity to be able to extend between 5 and 10 times higher than a flame that is not within a cavity, regardless of whether or not the surfaces of the cavity are combustible.

Regulation and BS 9991 require that the flame spread over or within an external wall construction should be controlled to avoid creating a route for rapid fire spread bypassing compartment floors or walls. This is an important consideration for any fire strategy but is of fundamental importance when a 'stay put' strategy is in place.

By utilising carefully selected vertical and horizontal cavity barrier products to sub divide and compartment concealed cavities, the rapid spread of fire from one compartment to another is prevented.

7.20.5 The location of Cavity Barriers

BS 9991 requires that cavity barriers should be provided in an external wall (Note: barriers are required in other locations as well):

- To close the edges of cavities, including around openings (such as window / door openings, extract vents, etc.).
- At junctions where an external wall cavity is in alignment with a compartment wall or compartment floor (noting that where a cavity wall is constructed with both the inner and the outer leaves being of masonry or concrete which are each a minimum of 75mm thick it does not need to comply with this requirement, refer to Figure 25 of BS 9991 for clarification).
- The important thing to remember is that we want to keep one compartment, separated from another to prevent a fire in one compartment spreading to another.

In addition BS 9999 requires that non-domestic* cavities, regardless of the locations of the compartment wall locations should not be allowed to extend beyond a certain dimension before additional cavity barriers are installed. Table 34 of BS 9999 should be referenced but for external walls a cavity that is lined:

- With products/surfaces which are National Class 0, BS 476 class 1 or are European class A1 through to C-S3, d2 the cavity should be subdivided every 20 metres.
- With products that are not any of the above classes should be subdivided every 10 metres.

For external walls these 'extensive cavity' requirements do not apply:

1. If a wall is required to be fire resisting only because it is load bearing, or
2. Where a cavity wall is constructed with both the inner and the outer leaves being of masonry or concrete which are each a minimum of 75mm thick, or
3. Where a cavity occurs behind the external skin of a cladding system which has a concrete or masonry inner leaf that is at least 75 mm thick, or is formed by over cladding an existing masonry (or concrete) external wall provided that the cavity does not contain combustible insulation and the building is not put to a residential or institutional use.

- Whilst BS 9999 requires that these requirements for additional cavity barriers are only applicable to extensive cavities of non-domestic buildings, it should be assumed that the requirements do also apply to residential high rise properties, advice and substantiation should be sought from a fire engineer prior to omitting these from properties of this kind.

7.20.6 Cavity Barrier types

Vertical cavity barriers will normally fully fill the cavity void and be compressed between the inner surface of the outer wall element and the outer surface of the inner wall element.

External wall cavities are often required to provide drainage or drainage and ventilation. When either of these requirements apply a horizontal cavity barrier that facilitates drainage and / or ventilation will need to be sourced, these are often provided with an intumescent strip to the front face or are fabricated from a perforated metal that is intumescent coated. The intumescent elements are activated during a fire by heat, and are designed to expand and close the cavity. These intumescent products need to be carefully selected to ensure that the intumescent is capable of closing and sealing the ventilation gap. The location of the cavity barrier needs to be carefully selected to ensure that the intumescent has a flat robust and continuous surface to seal against.

Note: Where both the inner and outer cavities leaves are of masonry or concrete that are 75mm thick or more (refer figure 25 of BS 9991) that the purpose of the cavity closures to the perimeters of openings is to restrict the airflow within the cavity, and therefore the closers may be of a material that might not conform to the recommendations in Table 3 (e.g. 30 minute integrity and 15 minute insulation), noting that BS 9991 19.2 states that every cavity barrier should provide 30 minute integrity as a minimum. Guidance and substantiation should be sought from a fire engineer before the fire insulation performance of these elements is reduced.

Clause 19.2 of BS9991 allows the door or window frame that is installed in an opening to act as the cavity barrier if the framing element is constructed from either 0.5mm thick steel or 38mm thick timber.

Cavity barrier suppliers should have a technical team who can advise on which product to use for any given position, and should be able to attend site to inspect the installation. Cavity barrier selections should be reviewed and approved by your fire engineer.

7.20.7 Installing and fixing Cavity Barriers

Every cavity barrier should be constructed and installed in such a way that it can provide the required level of performance. Wherever possible a cavity barrier should be mechanically fixed tightly in place to a rigid construction. When this is not possible it should be fitted in accordance with clause 24.4 of BS 9991.

When designing the cavity barrier installation the following items should be considered:

1. Building movements (settlement, shrinkage, thermal expansion, sway, deflections, etc.).
2. Collapse or deformation of structures or surfaces that the cavity barrier is attached to or to which they abut, or the services (ducts, pipes, etc.) that penetrate them.
3. Failure of the fixings due to fire.

Openings / penetrations in cavity barriers should be kept to a minimum. Any penetrations should be sealed to restrict the passage of smoke. In external walls all openings in a cavity barrier should be restricted and should comply with the constraints defined within clauses 19.2 and 24.4 of BS 9991.

7.20.8 Records

During the design and installation process, it is critically important that final as built and maintenance information is collated and included in the building fire safety manual.

BS 9999 W.3 requires that the records kept should include drawings which identify the position and specification of the cavity barriers as well as the locations of all fire compartment walls and floors; and that a certificate of completion should be obtained.

Copies of all records should be added to the fire safety manual.

Grateful thanks to Clive Everett
Façade Technical Standards Director
LABC Warranty

Statutory Requirements and Technical Standards

Approved document B – Fire Safety
BS 9991: 2015 – Fire Safety in the Design, Management and Use of Residential Buildings
BS 9999:2017- Fie safety in the Design, Management and Use of Buildings
BS 476:1997– Fire Test on Building Materials and Structures
MHCLG – BSP Advice note 16
BWF Certifire – Fire Doors and Doorsets. Best Practice Guide
LABC www.labc.co.uk
LABC Warranty www.labcwarranty.co.uk

Preparing for Handover

8.1 STRATEGY DEFINITION

Project handover involves the transfer of overall responsibility for the works from the construction team to the client's user organisation(s).

The handover strategy should be developed by the architect/employer's agent (A/EA) in conjunction with the client to suit the client's financial, operational and other requirements. This could include:

- phased handover, partial possession or sectional completion, releasing key areas before overall completion
- partial possession, enabling early access for specific activities, for example specialist fitting out, while the remainder of the construction work is completed
- early agreement of the items listed below.

It is good practice to establish the handover strategy early in the project as part of the detail design stage and to incorporate it into the master programme. However, there is enormous variation between projects.

Certain elements may be addressed in the specification, such as training provision by the contractor's installers for the client's maintenance team on safe operation and maintenance aspects.

As the detailed design develops, the handover strategy should develop into a more detailed handover plan. This could cover:

- project completion
- timetable for handover and lead-in activities
- responsibilities during the process
- training requirements
- handover procedures (including to non-client organisations)
- involvement of maintenance staff/organisations
- demonstrations and witnessing requirements
- any 'familiarisation' programmes required
- commissioning plans with appropriate method statements
- lift tests
- fitting out
- documentation, including the Health and Safety File
- environmental impact information
- advice to the client (e.g. on transfer of insurance liability)
- handing over of keys from the main contractor
- operation and maintenance/building manuals, and a statement of the person authorised to keep them and their place of storage; a draft copy should be prepared so that those who need to can see what is in it.

After the plan has been developed with the design team and the client, the client should formally approve it.

8.2 CLERK OF WORKS' AND SITE INSPECTOR'S ROLES

The Clerk of Works will be expected to monitor the procedures and actions identified in the commissioning and handover plans. The A/EA or services engineer may require the inspector to witness appropriate tests. Some inspections may need to be carried out or witnessed by external parties, such as insurance inspectors, or fire or safety officers. The client's maintenance organisation may also wish to be present. During commissioning, all appropriate health and safety measures should be in accordance with the method statement attached to the commissioning plan.

These measures may include permit to work areas or isolation of certain equipment during testing.

Where self-certification is legitimately carried out by subcontractors, the A/EA will need to verify that all requirements for certification have been clearly allocated to contracts, so that the Health and Safety File and the Operation and Maintenance/ Building Manual will be complete. The A/EA may be required to assist and monitor delivery of the testing and signing certificate.

Where appropriate, the Clerk of Works should liaise with the health and safety coordinator to establish that test certificates for service installations are being obtained and incorporated into the Health and Safety File.

8.3 DEFECTS AND DEFICIENCY (SNAG) LISTS

It is generally accepted good practice to try to keep the 'snag' list to a minimum by encouraging the contractor to take action to remedy faults well in advance of the final handover or occupation date, whichever is earlier.

The Clerk of Works should carry out a thorough preliminary inspection with the contractor, record all defects and omissions and, after agreeing them with the A/EA, issue instructions that they are to be put right before the date of handover where possible or by a given date. A long initial snag list should indicate that the work is not ready for inspection. In a phased handover, such as a housing project, items picked up in the initial snagging should not be expected to keep reappearing.

It is increasingly common to use computer-aided defects tracking systems on sites. These can be used to log defects and clearance dates, sorting them into trades and locations if required. The re-sort facility is particularly useful, as the lists can be prepared on a room-by-room basis and then sorted into separate lists for subcontractors.

The keys to successful snag lists are method and consistency. All inspectors have their own ways of achieving this; for example, in a building of two or more storeys, starting externally from the roof, then each elevation with all its elements clockwise. Then starting internally from the top floor and working down – on entering a room, checking for defects on the ceiling first, then the walls in a clockwise direction starting from the entrance door, including windows and any stores off the room, any fixed furniture and finally the floor, before checking the door and exiting. Finally, walk clockwise around the unit/property to list any external snags.

It is the responsibility of the Clerk of Works to verify that the snag list is being cleared (agreeing division of building, civil engineering, M&E and landscape responsibilities), and to carry out the final snagging inspection, ensuring that all special tools, specified spares, fire appliances and notices are in place. Building control may have separate requirements (i.e. public licensing issues), and the Clerk of Works may well be involved in monitoring these to completion.

8.4 WITNESSING AND TESTING: PROCEDURES GENERALLY

Clerks of Works should try to ensure that adequate notice is given to allow them to arrange workloads to suit, and that a method statement is available. It is not uncommon for the commissioning programme to be telescoped because the contract is delayed, leaving an impossible witnessing programme. It is wise to alert the architect as soon as an indication of this kind of problem becomes apparent.

It is the duty of the Clerk of Works to verify that all test instruments have been calibrated and are accompanied by a current certificate.

Buildings or facilities must be in a fit state for the test and represent acceptable conditions; for example, it is not acceptable to carry out heating or ventilation tests if doors are propped open for other trades to work.

Where tests are to be carried out by a 'competent person' (see section 8.4.1), the Clerk of Works must ensure that the contractor is aware of the requirement. It is also suggested that if the competent person is not a representative of a company known to the Clerk of Works as practitioners in this specialised field, a company brochure and list of clients for whom they have carried out similar work should be requested before testing and inspection proceeds. The architect should be alerted if any uncertainty remains about the competency of the competent person.

8.4.1 Witnessing generally

It is good practice to confirm early in the project with the client and/or the architect the extent of witnessing required, and to confirm the amount of time available to be spent on this activity. Work will have to be prioritised; for example, there is little need for a Clerk of Works to be present all the time for inspections carried out by a competent person or independent specialist if both the A/EA and the Clerk of Works are happy with the method statement and their initial activities on site. However, if any dissatisfaction remains with the procedures or quality of any testing and commissioning, it must be reported to the A/EA or services engineer and the problem discussed.

Full site diary records of all timing and commissioning activities must be kept, including when competent people and other specialists are on site.

8.5 HANDOVER DOCUMENTATION

The Construction (Design and Management) Regulations 1994 (or as amended by the CDM Regulations 2006) set out the requirements of handover documents (refer to the *CDM Regulations Approved Code of Practice*). In practice, the needs will be defined in collaboration with the health and safety coordinator.

8.6 AS-INSTALLED OR AS-BUILT DRAWINGS

The Clerk of Works should monitor progress on (and accuracy of) the contractor's as-installed drawings and the compilation of data for the Operation and Maintenance/Building Manual as directed by the A/EA and the Health and Safety File as directed by the health and safety coordinator.

8.7 COMMISSIONING

Commissioning involves a great deal more than just the services: lifts, the wet risers, air conditioning, etc. On the construction side, windows, roofs and 'person-safe' systems all need commissioning as well, often for insurance purposes, but the following sections all deal with the services element.

8.7.1 Witnessing

The Clerk of Works' role here may include:

- witnessing final high voltage (HV) and other electrical tests, and obtaining and countersigning the test certificates
- verifying that all pre-testing and pre-commissioning cleaning of pipe work and ductwork is carried out
- witnessing all pressure tests on pipe work and ductwork leakage tests where applicable
- witnessing and assisting the insurance and fire officers for certificate testing, if applicable, on behalf of the client.

If applicable, the Clerk of Works may have a responsibility to verify that arrangements are made for the initial delivery of solid and liquid fuels in time for plant start-up.

8.7.2 Plant commissioning

- Before use, calibration of any instruments to be used for commissioning tasks must be verified.
- As agreed with the M&E project engineer and A/EA, the Clerk of Works must witness that air and water distribution systems are correctly balanced, checking this at measuring stations confirmed with the M&E project engineer. The commissioning of all plant, e.g. boilers, pumps, fans, compressors, coolers, chillers, air handling units, water treatment and control and indication systems must be witnessed. The operation of all safety devices must be verified by simulating conditions.
- The Chartered Institute of Building Services Engineers (CIBSE) Guide contains further useful information. See Chapter 17.
- The Clerk of Works should liaise with specialists appointed for the testing of air conditioning for hospital operating theatres, computer suites and process systems; intruder detection systems; noise, vibration and contamination control in nuclear, biological or chemical protected buildings; and any other highly specialised systems.
- Any remaining site M&E proving and performance tests must be witnessed to ensure that certification is correctly completed. The project M&E engineer and others concerned should be notified immediately if anything arises that is likely to jeopardise the handover arrangements.

8.8 PLANNED MAINTENANCE

If requested by the A/EA, the Clerk of Works should verify that the planned maintenance detail sheets have been completed in time for their inclusion in the operation and maintenance manuals before handover.

8.8.1 Operation and maintenance/building manuals

It is one of the duties of the Clerk of Works to collect, as agreed with the building project manager and M&E project engineer, documentation for the A/EA to collate into the manual and to ensure that all M&E manufacturers'/contractors' operating and maintenance instructions, literature and test certificates are provided, including:

- the actual performance data obtained from commissioning
- the schedule of M&E equipment
- the planned maintenance log book
- any risk assessments and Health and Safety Plans
- the fire alarm and emergency lighting inspection and test certificates required under BS 5839-1:2013 and BS 5266.

The Clerk of Works will assist the M&E project engineer to ensure that the operation and maintenance manuals are adequate.

8.9 SUMMARY OF TEST CERTIFICATES

The following table gives a summary of the main categories of tests that may need to be witnessed and the certificates to be included in the operation and maintenance manuals/Health and Safety File.

	Item	Source
1	Electrical installation certificate	Tests carried out by the electrical contractor and witnessed by the Clerk of Works Results recorded on agreed forms based on the model forms in the IET Wiring Regulations BS 7671
2	Earthing, lightning protection, anti-static precautions, etc., test certificates	Tests carried out by the electrical contractor and witnessed by the Clerk of Works Results recorded on agreed forms
3	Lifting equipment test certificates	Architect to raise order on competent person (CP) Tests carried out and certificates issued by CP
4	Lifts and hoists (and lifting beam) test certificates	Tests carried out and certificates issued by CP
5	Steel wire ropes test certificates	Certificates issued by the rope manufacturer
6	External heating mains test and commissioning certificates	The contractor, or the commissioning specialist approved by the architect, is responsible for showing that the installation meets with the design criteria
7	Heating and domestic hot water installations test and commissioning certificates	Off-site tests for some items of equipment are required. These are carried out at the manufacturer's works, where the test certificates are issued, e.g. for boilers, calorifiers, pumps, fans, motors, starters and control gear The contractor, or the commissioning specialist approved by the architect, is responsible for fully commissioning each installation as specified and for issuing the commissioning certificate

	Item	Source
8	Gas, air, water, etc., part system pressure test, test certificate	Test carried out by the contractor according to the specification, witnessed and test certificate countersigned by the Clerk of Works
9	External gas mains and internal gas installations test certificates	Distribution mains external to the site and service pipes within the site up to and including the primary meter are the responsibility of British Gas/Transco Certificates for the installation beyond that point are the responsibility of the contractor
10	Mechanical ventilation and air conditioning test and commissioning certificates	Off-site tests for some items of equipment are required at the manufacturer's works. The manufacturer will be responsible for the issue of the test certificates, e.g. fans, pumps, electric motors, starters and control gear, air heaters, refrigeration plant The contractor is responsible for fully commissioning each installation, issuing the commissioning certificate for ventilation and air conditioning
11	Pressure vessels test certificates and reports	The certificates are issued by the CP The architect is to arrange for the CP (for boilers and pressure vessels) to undertake: ● approval of design, examination under construction and test at the manufacturer's works, and issue of a certificate of test unless these services have been arranged by the manufacturer ● examination and test of second-hand plant under purchase ● initial examination and test after installation on site and issue of report forms prescribed under the Factories Act ● examinations of test welds where required
12	Safety officer's report	Where advised necessary by the A/EA, the report is completed by the client's safety officer

	Item	Source
13	Building control certificate	Within design and build contracts the contractor is responsible for obtaining this. The Clerk of Works should make sure it goes into the operation and maintenance manuals

Handover, Practical Completion and Defects Rectification Stage

9

9.1 HANDOVER: GENERAL PROCEDURES

The takeover of completed works from the contractor, and the acceptance from the contractor by the client for subsequent maintenance, is a contractual event.

Handovers need not involve the whole site; they may be of only one part of it, and this is often known as phased, partial or sectional completion. The A/EA will normally give directions as to when the handover will occur and any specific procedures to be followed, but the Clerk of Works will often be asked to provide advice and reports on the state of the works. The A/EA will need to liaise with the client and any maintenance organisation beforehand to ensure readiness, and to make certain that any difficulties or problems are resolved before formal handover to the client.

Detailed procedures for takeover and handover are to be agreed with the A/EA. These will depend on the particular requirements of the client, and the A/EA is responsible for making the necessary arrangements.

Generally, not less than four weeks before the probable date of handover to the client of any part of the works, and eight weeks or more in advance in the case of a complete installation, the A/EA should notify the client with a request to make preliminary arrangements to take over the building or other works services. When occupation or partial occupation of a building is to be taken before actual completion of the contract, the A/EA will also give notice so that arrangements for the client's maintenance and changeover of building insurances can be made.

As soon as possible after notice is given, a site meeting should be held, attended by the A/EA, client and maintenance organisation or their representatives, to discuss any outstanding items and to agree the takeover date.

The Health and Safety File will require a schedule of principal materials used in construction. The coordinator and the architect will determine how this is to be produced, and may require the site Clerk of Works to assist in the preparation. The Clerk of Works should verify that the architect has been provided with all test reports and certificates, prior to the practical completion date. The contractor is ordinarily responsible for provision of as-built drawings, although the consultants may prepare part of them. In either case, these should be checked against the Clerk of Works' own record of variations and from site knowledge.

9.1.1 Defects and deficiencies (snag lists)

Depending on the form of contract, a list of outstanding defects will be compiled as the project approaches handover. The A/EA and the Clerk of Works, supported by the design consultants, should inspect the works and compile a list suitable to be attached to any practical completion certificate. It is advisable to verify whether the contractor should also be in attendance.

9.1.2 Training of the client's maintenance team

Where required, the training of the client's/user's maintenance organisation should be arranged to achieve the requirements of the handover plan, including any installation or manufacturers' instructions.

9.1.3 Health and safety file and as-installed drawings

All information necessary for the end user to maintain and operate the final works is to be incorporated into the Health and Safety File. This is reviewed by the coordinator before issue to the client.

It is good practice to monitor progress on (and accuracy of) the contractor's as-installed drawings or the last issue of contract drawings 'for construction'. This will include the compilation of data for the operation and maintenance manuals and Health and Safety File as required by the coordinator.

9.1.4 Pre-completion planning meeting

A pre-project completion meeting should be held at a suitable time before anticipated project completion (generally this will be approximately two weeks beforehand).

The meeting is normally attended by the client and consultants, and may include deliberation on the following:

- a review of the quality being delivered and the level of defects
- performance to date and expectation of completion date
- state of appearance and cleanliness
- extent of incomplete works
- progress on services commissioning and testing
- Health and Safety File documentation
- damage
- readiness of the client's organisation to take on the building and consequent responsibilities (e.g. day-to-day maintenance, running costs and insurance)
- certification from the building control officer.

The purpose of the meeting is to review the standards of construction achieved and all outstanding or incomplete works, including documentation for the Health and Safety File. From this review, the likely date of project completion is established, and any further actions agreed.

9.1.5 Keys

The Clerk of Works may be asked to verify availability of keys immediately before handover with the contractor and to ensure that the keys are clearly labelled. This does not just apply to door keys, but also to window, manhole and any engineering (services) keys. If for any reason the Clerk of Works is obliged to receive important keys significantly in advance of handover (e.g. because of staged completion), the architect should provide a security key case and carry out appropriate security measures, perhaps with the client's security advisor. The contractor may also have a key release form.

The contractor is likely to require a receipt for any keys handed over; this is normally the architect's responsibility unless delegated to the Clerk of Works.

Where special security locks and keys are specified (security lock keys and master keys should not normally be consigned to the contractor), there may be specific requirements for installation and protection of high security locks before, during and after installation; the Clerk of Works should verify the project specification and with the A/EA.

In any event, any key handed over to a client or client's representative must be signed for.

9.1.6 Operation and maintenance of the finished building and maintenance documents

It is likely that maintenance documents specified in the contract, including those provided by suppliers and subcontractors, will go straight into the operation and maintenance manuals. It is also good practice to see that these are completed in the prescribed format and are ready for handing over to the client on the due date. The client's cleaning contractors may require early access, for example if there is any special flooring.

The Clerk of Works may be asked by the coordinator to verify that all information required for the Health and Safety File has been prepared and issued by the designers, contractors and any other parties.

9.1.7 Handover meeting

It is normal practice for the client and consultants to attend the project handover meeting. If everybody is satisfied that construction work is practically complete and documentation is in an acceptable state, then the practical/partial completion certificate can be signed by the A/EA. Generally appended to the certificate is a complete list of outstanding works and defects. This is to be agreed by all the parties involved. However, it should be made clear that this does not alter the contractor's responsibility for remedying any defects identified after project practical completion, within the specified defect rectification period.

The A/EA is normally responsible for creating the practical completion certificate. Forms produced by the respective professional institute (e.g. RIBA) are often used. The Clerk of Works' role is normally confined to the production of an up-to-date schedule of outstanding works and defects and deficiencies (the snag list), which is attached to the certificate.

Once project completion has been certified, no additional works can be instructed by the client, the A/EA or the Clerk of Works under the original contract. The project completion certificate is issued to the contractor, with copies to the client and all the project consultants. A copy should be retained on the inspector's site files and the end of the rectification period (defects liability period) recorded in the site diary.

However, if at the meeting it is agreed that the works are not in a fit state to hand over, for whatever reason, the architect has the authority to issue a non-completion certificate. The Clerk of Works may be asked to assist the architect to record the reasons for its issue.

Handover of the project should not proceed until project/practical completion has been achieved. Any other strategy carries significant risk. Once this has taken place the client controls the project and becomes responsible for security, building insurances, day-to-day maintenance and operation. Access for defects rectification must be coordinated through the client's nominated representative.

Arrangements are made for clearance of defects as they occur throughout the defects rectification period. The original project team generally makes these arrangements. The Clerk of Works may be asked to verify and agree any areas of damage made by client's removal operators or from 'white goods' deliveries.

9.2 CONDITION AT COMPLETION

The contractor is normally required to remove all temporary markings, coverings or protection unless otherwise instructed, and to clean and clear the works of debris and rubbish, leaving the premises on completion in a fit condition for occupation and use. If any special degree of cleanliness is required, for example in technical or prestige areas, this should be detailed in the specification or otherwise ordered. The contractor's attention should be drawn to these matters in the course of the defects and deficiencies inspections.

9.3 PARTIAL POSSESSION AND SECTIONAL COMPLETION

If it becomes evident that dates specified in the contract for obtaining partial possession of sections of the site or for completing and commissioning engineering services in advance of the main contract are unlikely to be met, the A/EA must be informed as soon as possible.

9.4 POST-HANDOVER AND RECTIFICATION (DEFECTS LIABILITY) PERIOD

This period may have other titles, such as the rather misleading 'maintenance period' for government contracts.

The period runs from practical completion (i.e. the date on the practical completion certificate) for the period specified in the contract. This is usually six or 12 months. Twelve months is a desirable period for M&E installations so that a complete annual heating/cooling cycle is experienced.

During this period, any residual works and defects should be completed by the contractor. There is also the possibility that other defects may come to light during the first year of occupancy, for which the contractor is also responsible.

The Clerk of Works may be involved in:

- monitoring the completion of items known about at handover and recorded on the snag list
- inspecting and advising on defects that come to light during this period (there may be a dispute over whether the defect was caused by the occupier or is the responsibility of the contractor), and adding these to the snag list when instructed by the A/EA
- advising on events that have led up to contractual disputes, such as claims for loss, expense, disruption and extensions of time
- completing and filing records
- attending any post-completion workshops/reviews that occur, for an insight into the good and bad elements of the project.

9.5 RECORD DOCUMENTATION

On completion of the contract the Clerk of Works should seek instructions from the client regarding the return of all drawings, specifications, bills of quantities and other contract documents, the Clerk of Works' own copies of instructions, a copy of the contractor's programme showing actual progress made and other formal records. Especially valuable records include relevant photographs, site diaries, daily labour returns and the Clerk of Works' dimension book, giving measurements of works covered up, for example on additional foundation excavations. Where the A/EA is an external appointment, all records should be passed to the client.

The Clerk of Works' site records may often be the most authoritative contemporaneous evidence of disputes between the contractor, the design team, the A/EA and the client. Included in these records is the Clerk of Works' marked-up drainage drawing, often a most useful record of the revisions that actually occurred during construction.

9.6 SURPLUS MATERIALS AND EQUIPMENT

The A/EA may identify surpluses that are client property and not part of the 'spares' list, and issue directions for disposal.

9.7 PREPARATION OF DEFECTS LIST DURING THE RECTIFICATION (DEFECTS LIABILITY) PERIOD

When repairs are required during the rectification period, the A/EA should determine whether they are the contractor's responsibility and, if so, advise the contractor so defects can be rectified as soon as possible within an agreed time frame. If the Clerk of Works has reason to believe that the contractor is not keeping up or making sufficient progress, the A/EA should be informed immediately. The reporting of defects should not be left until the end of the rectification period(s) – which, it should be noted, may vary for building and M&E works.

Any additional defects, reported from any source including the client/client's agent should, if appropriate, be reported to the A/EA before being added to the snagging database/list.

9.8 CONTRACTUAL CLAIMS AND DISPUTE RESOLUTION

For claims and counter-claims, the A/EA must consider all the possible implications and consequences of the potential and issued instructions. As early as possible, steps should have been taken to detect any circumstances that might develop and give rise to additional expense or claims. Consideration should also be given to ensuring that adequate contemporaneous records have been kept and maintained in order to ascertain expenses and damages. The A/EA should liaise closely with the quantity surveyor on all these matters.

Claims for prolongation and disruption expenses arising in the circumstances provided for in the contract or stemming from unreasonable delay by the project team in issuing or approving drawings that the contractor is required to submit, and extra-contractual claims, will ordinarily be negotiated by the quantity surveyor.

The A/EA must advise the client of any claims received together with the A/EA's comments and recommendations, and the Clerk of Works may be asked to give an account of events and demonstrate site records. The site diary is often the most complete documentary evidence of the events leading up to the claim situation.

The quantity surveyor may require the assistance of the Clerk of Works to settle any extra-contractual claims.

The client should be kept informed of developments, if the client has instructed the Clerk of Works to report directly, rather than through the A/EA.

It should be noted that resolution of claims would normally attract an additional fee or time charge. The Clerk of Works should seek agreement from the client before committing resources.

9.9 INSPECTION AT EXPIRY OF RECTIFICATION (DEFECTS LIABILITY) PERIOD

On expiry of the rectification period (or before, if possible), the Clerk of Works should verify that the contractor has completed all outstanding work, and has made good all defects listed on the practical completion certificate and all defects subsequently reported during the rectification period.

Once satisfied that all outstanding work is completed and all defects have been made good, the A/EA will issue the certificate of making good to the contractor, also sending copies to the client, quantity surveyor and Clerk of Works.

9.10 PERFORMANCE REPORTS AND PROJECT FEEDBACK

It is good practice for the Clerk of Works to assist the A/EA to complete performance reports on the contractor and subcontractors as required by the client.

The Clerk of Works should provide feedback to the client on the project generally, including on any design-related issues, and request feedback accordingly. Any matters that could prevent problems recurring on future projects should be identified. Feedback should be given at any time during the contract, not only at completion.

Typical performance feedback may include information on:

- suppliers
- contractor
- subcontractors (domestic or otherwise)
- materials used
- the Clerk of Works' own performance
- consultants
- client staff
- design details
- training needs
- key performance indicators (KPIs).

The Clerk of Works may be asked to attend the client's post-completion workshops/reviews (usually held 18 months after practical completion) for an insight into the good and bad elements of the project, and its aftercare.

Published and Model Forms for Clerks of Works

10

10.1 SPECIMEN CLERK OF WORKS PROJECT REPORT

10.1.1 General

The project report included within this chapter has served well for many years. It also provides an excellent pro-forma on which to base a bespoke template in an electronic format, which may better suit current site administration protocols and expectations.

The main purpose of the weekly report is to inform the design team and other relevant stakeholders of the status of a project at a given point in time – usually the end of the previous week's site activities. However, for the report to be of value, the Clerk of Works must ensure both impartiality and currency. Consequently, the report content needs to be:

- accurate
- demonstrably factual (not including or reliant on unsubstantiated opinion)
- relevant
- unambiguous
- impartial
- concise
- impersonal
- timely.

10.1.2 Notes on use and completion

For the purpose of these notes, it is assumed that a JCT form of building contract is being used. However, the project report form can be used under other forms of contract.

1. The name of the project, which will be its contractual title, and its address, i.e. the address of the works or site, should be entered. These descriptions should be maintained consistently throughout all forms and records relating to the project. The project reference number is given to the project for administrative purposes.

   The report number (which should be consecutive) and the week-ending date should be entered. This will be the Sunday at the end of that week. If project reports are to be less frequent than weekly, the most convenient approach will be to complete the form as appropriate and attach separate weekly sheets as required for trades on site, weather, etc.

   The name of the contract administrator (CA) and the name of the contractor, all as shown in the building contract, should be entered. The Clerk of Works' own name, together with contact details for site, should be entered.

The contract start date should be the date stated in the building contract for possession or commencement, as appropriate. The contract completion date should be the date stated in the building contract or as last revised following any extension of time.

When required, the progress against programme entry should indicate where, in the Clerk of Works' view, slippage has occurred or the main contractor is ahead of programme. The entry may be expressed in days or weeks, or as a percentage deviation from the current programme issued by the main contractor.

2. The number of person-days per trade for each day of the week should be entered. For example, four operatives working all day will be entered as four person-days (assuming a working day is seven hours on site). Two operatives both working all morning only will be one person-day. Two operatives working one in the morning and one in the afternoon will also be one person-day. Enter the total number of person-days for each day of the week.

Frequently, on large projects, labour recording relies on biometric recognition systems. If agreed with the CA, this panel may be left blank and the main contractor's labour return in electronic format substituted or attached in addition. An important purpose of these entries is to enable the CA to verify that the main contractor has been proceeding regularly and diligently. Care is therefore needed to make sure that entries for the appropriate categories are consistent, week on week, and that operatives are not accidentally included twice (e.g. as machine operators and as ground workers).

It may also be helpful to identify which trade operatives are working (for either a nominated or named subcontractor). This information will be invaluable for an A/EA reviewing extensions of time or direct loss and expense applications relating to nominated subcontractors.

3. The weather report entries should include a temperature reading. Ideally, this will be taken at the same time each morning and afternoon, otherwise a rise and fall reading of minimum and maximum temperatures can be given. In normal conditions, entries under a.m. and p.m. can be expressed as standard weather report headings (e.g. rain, bright, cloudy, thunder, sunny, dull). However, to alert the CA to weather conditions that might affect the main contractor's performance it might be expedient to agree to make the record under limited headings (i.e. snow, frost, rain, high winds, hot, dry), rating them on, say, a 1 to 5 scale that might assist computer analysis. Whatever approach is adopted, consistency of expression is important.

In the case of high-rise construction, weather conditions at ground level can be markedly different from those at high-level elevations. This should be borne in mind and noted as appropriate.

Person-hours lost due to adverse weather conditions should be given for each day, with the weekly total and total to date recorded.

4. All visitors to site should be noted, with names, functions and dates.

5. Under general comments the Clerk of Works might draw the CA's attention to matters that require immediate action, for example a boundary dispute with the owners of a neighbouring property.

6. The entry for site directions should include Clerk of Works directions.

7. Delays or events with potential to cause delay should be listed. The latter are often expressions of opinion by a vigilant Clerk of Works, but they will be invaluable when the CA is considering claims for extensions of time.

8. Defective work observed, including instances of non-compliance with the building contract in respect of workmanship, materials, working methods, etc., should be recorded, together with a note of any action taken by the main contractor to remedy such deficiencies.

9. Drawings/information received on site should include any CA instructions and information issued by the main contractor (e.g. method statements) that might have been requested. Also included might be relevant notices or explanatory leaflets received from the Health and Safety Executive.

10. Drawings/information required on site by the Clerk of Works should include items already requested but not so far received.

11. The entry under plant/materials should include items delivered to site during the week, and measures for their protection as required by the building contract. The removal from site of any unfixed materials and goods should be noted, and whether this was done with the consent of the CA as required by the building contract.

12. The progress to date panels include items or operations described in a way that is generally accepted within the building industry. The percentage entries indicate the percentage of work actually achieved as measured against the percentage of work that should have been achieved according to the main contractor's current programme. For example, 'by the end of that week 70 per cent of the roof covering should have been completed, but in fact only 60 per cent has been carried out. There is therefore a 10 per cent shortfall against programme for that operation'.

13. The general report will summarise the contents of the Clerk of Works' diary for that week and would include recording issues relating to workmanship and materials, any tests witnessed, day works notified, third-party certifications received, or relevant events noted. Where NEC forms of contract are used, early warnings and compensation events should also be highlighted. This information might be

invaluable to the CA when dealing with extensions of time and other contractual issues. Both the reports and the site diary entries might subsequently be used as evidence in arbitration or litigation proceedings, and Clerks of Works should bear this in mind when compiling and setting out facts and observations.

14. Photographs are an excellent medium for supplementing information and can be easily imported into a report in digital format. Photographs should be dated, annotated and located to a position on site. A prudent Clerk of Works will ensure that they capture significant issues, such as the positioning of cavity barriers or fire-stopping before void closure, etc.

15. The health and safety entry might draw the CA's attention to instances of non-compliance with the regulations (e.g. in respect of safety clothing or scaffolding) or work not in accordance with the method statement or safety plan. Any action taken by the Clerk of Works or intervention by the HSE Inspectorate should be noted.

16. The entry under site housekeeping will describe conditions on site during that week. For example, the site might have become a quagmire because of continuous heavy rain, making working conditions extremely arduous. Observations of this kind will help the CA when dealing with notices of delay.

17. Enclosures might include the main contractor's labour returns (see point 2, above), site directions issued by the Clerk of Works and reports. The latter might concern the weather, air quality, cube tests, commissioning, performance of components.

18. The distribution of the project report should be agreed with the CA, to whom the top copy should be sent.

19. The Clerk of Works whose name is given on the face of the form should sign and date the project report.

CLERKS OF WORKS PROJECT REPORT

PROJECT:	REF:
ADDRESS:	NO:

Architect/Contract Administrator:	Week/Month ending:
	Contract Start Date:
Main Contractor:	Contract Completion Date:
Clerk of Works Phone/Fax No:	Progress +/- to Programme:

TRADES	MON	TUES	WED	THU	FRI	SAT	SUN
Site Staff							
Groundworkers							
Steelfixers							
Steel Erectors							
Concretors							
Drainlayers							
Machine Operators							
Carpenters							
Scaffolders							
Bricklayers							
Roof Finishers							
Wall Cladding							
Window Fixers							
Glaziers							
Floor Screeders							
Plasterers							
Tilers – Wall/Floor							
Dryliners/Partitions							
Ceiling Fixers							
Decorators							
Floor Finishers							
Heating/Ventilation							
Plumbing							
Electricians							
Hard/Soft Landscape							
Roadworks							
Public Services							
TOTAL							

SITE DIRECTIONS ISSUED

No	Item		Date

DELAYS INCLUDING DEFECTIVE WORK (ACTION TAKEN)

DRAWINGS/INFORMATION RECEIVED ON SITE

DRAWINGS/INFORMATION REQUIRED ON SITE

PLANT/MATERIALS DELIVERED TO SITE OR REMOVED

WORKS – MATERIALS: TESTED OR COMMISSIONED

WEATHER REPORT

	AM	°C	PM	°C	Time Lost
Mon					
Tue					
Wed					
Thur					
Fri					
Sat					
Sun					

Stoppages (Hours)	
Total to Date	

Visitors (include Statutory Inspectors)	Date

	Progress to date	Programme		Progress to date	Programme		Progress to date	Programme
	%	%		%	%		%	%
Preliminaries			Blockwork Internal			Electrical 2nd Fix		
Excavation			Cladding/Curtain Wall			H & V 2nd Fix		
Shutter/Reinf			Windows-glazing			Ceiling Grid/Tiles		
Concrete Structure			Joinery 1st Fix			Decoration		
Steel Erection			Plastering			External Works		
Main Drainage m/h			Drylining/Partitions			Hard/Soft Landscape		
Floor Construction			Floor Screeds			Roadworks		
Floors Suspended			Plumbing 1st Fix			Mains: Gas-Electrical		
Roof Structure			Electrical 1st Fix			Mains: Water-Telecoms		
Roof Coverings			H & V 1st Fix			Lifts		
Drainage fw.sw.			Wall/Floor Tiles			Alarm/Computer Systems		
Brickwork External			Plumbing 2nd Fix			Defects/Handover		

GENERAL REPORT: SUMMARY OF WORKS IN PROGRESS

HEALTH AND SAFTETY: ACTION TAKEN

SITE HOUSEKEEPING: CONDITIONS/CLEANLINESS: ACCESS AND EGRESS

Enclosures: G.C. Labour Return [ ] Site Directions [ ] Reports (State) [ ]	OFFICE ACTION:
Distribution as Agreed	
Client [ ] Architect [ ] Project Manager [ ] Quantity Surveyor [ ] Office [ ] Others [ ]	
Clerk of Works Date	

10.2 SPECIMEN MAIN CONTRACTOR'S LABOUR RETURN

10.2.1 Notes on use and completion

To be completed by the main contractor and submitted weekly to the Clerk of Works.

1. The name of the project and its address should be entered and should be consistent with the Clerks of Works' report. The week number, as shown on the programme for the works, should be entered together with the date of the Sunday at the end of that week. The project reference number will be the unique number given to the project for administrative purposes.

2. The number of person-days per trade for each day of the week should be entered. For example, four operatives working all day will be entered as four person-days. Two operatives, one working in the morning and one in the afternoon, will be one person-day.

3. A list should be entered of the main contractor's resources on site during that week. All heavy plant and specialist equipment should be included, but not standard minor items such as hand tools. Any equipment on site that has been hired by the main contractor should be separately identified.

4. Entries should clarify which materials and goods have been ordered and delivered in the course of the week, and which are still awaiting delivery.

Main Contractor's **Labour Return**

Project: _____   Ref. No: _____
Address: _____   Week No: _____
_____   Week ending: _____

ICWCI
Founded 1882

TRADES (Man days)		Mon	Tue	Wed	Thu	Fri	Sat	Sun
	Person in charge							
	Scaffolders							
	Groundworkers							
	Drivers and plant operators							
	Steel fixers and concreters							
	Bricklayers and masons							
	Roofers: Finishes							
	Carpenters and joiners							
	Steel erectors							
	Window fixers							
	Plumbers							
	H & V engineers							
	Electricians							
	Lift engineers							
	Plasterers and screeders							
	Tilers: Wall and floor							
	Ceiling fixers							
	Floor finishers							
	Glaziers							
	Painters							
	Drainlayers							
	Statutory undertakers							
	Road pavers, Tarmac workers							
	Total man days							

PLANT ON SITE

_____ _____ _____
_____ _____ _____
_____ _____ _____
_____ _____ _____
_____ _____ _____
_____ _____ _____

MATERIALS AND GOODS   Ordered and awaiting delivery

_____
_____
_____

Delivered to site

_____
_____
_____
_____
_____
_____

Signed _____   (for Main Contractor)   Date _____

This Return is to be completed by the main Contractor where this is required under the contract conditions, and submitted weekly to the Clerk of Works.

10.3 SPECIMEN CLERK OF WORKS DIRECTION FORM

10.3.1 Notes on use and completion

Directions may be given to the contractor by the Clerk of Works in respect of matters that are empowered under the building contract, and which may be the subject of a CA's instruction. A direction will be of no contractual effect unless confirmed by a CA's instruction within two working days. Any action before such confirmation will be premature and taken at risk.

The intention of a Clerk of Works direction is to alert the CA to information or action that, in the opinion of the Clerk of Works, is relevant at this time.

1. The name and address of the CA, employer and contractor should be completed as named in the building contract.

2. The works should be identified (i.e. the site address) and the date of the contract, as entered in the building contract.

3. The job reference, sequential number of the direction and date of issue should be given.

4. The content of the direction should be expressed clearly and concisely, itemising matters as appropriate. Care should be taken that numbers and dimensions are accurate and legible.

5. The declaration should be signed by the Clerk of Works.

6. The original should be issued to the contractor and, at the same time, a copy sent to the CA.

Clerk of Works Direction

Architect/CA:
address:

Employer:
address:

Contractor:
address:

Works:
situated at:

Contract dated:

Job reference:

Direction no:

Issue date:

Under the terms of the above-mentioned Contract, I issue the following direction.

This direction shall be of no effect unless confirmed in writing by the Architect/Contract Administrator within 2 working days, and does not authorise any extra payment.

Direction	Architect/CA use
	Covered by Instruction no:

Signed _____ Clerk of Works

Distribution	☐ Contractor	☐ Architect/CA	☐	☐ Site records file

F510 for SBC

RIBA 🏛 Reproduced by agreement with the RIBA

The Clerk of Works/NEC Supervisor: Responsibilities Within Standard Building Contracts

11

The main contract used for construction in past years has been the JCT Standard Building Contract, with the most recent version, the 2011 suite, taking account of changes to the housing grants, construction and Regeneration Act 1996. This contract is still readily available and the Clerk of Works is still named within it.

Many clients now look for an alternative, and the NEC family of contracts is coming to the fore, with the following characteristics.

- Its use stimulates good management of the relationship between the two parties to the contract and, hence, the works involved.
- It can be used in a wide variety of commercial situations, for a wide variety of types of work and in any location.

The Clerk of Works/construction inspector is identified as the supervisor.

The Institute of Civil Engineers (ICE) promote Clerks of Works as prime candidates to carry out the role of the supervisor, and they offer training to enable those who pass the required standard to be eligible for work from a register of supervisors.

11.1 ARCHITECT'S DUTIES

The architect under the terms of the RIBA architect's appointment within the JCT contract is required to 'administer the building contract, including reviewing the progress of construction works against the contractor's programme'. The project manager oversees works on behalf of the client within the NEC contract, and they may delegate additional responsibility to the supervisor.

The architect is not, therefore, required to make frequent or constant inspections and, where these are required, a Clerk of Works should be employed by the employer, who has such an entitlement under the JCT standard form of contract. The designer (architect) within the NEC is appointed by the client (employer) and is not referred to in the contract between the employer and the contractor.

11.2 CLERK OF WORKS' REMIT

Under clause 3.4 of SBC05, the Clerk of Works' duty is 'to act solely as inspector on behalf of the Employer under the directions of the Architect/Contract Administrator'. This clause also states that any Clerk of Works direction 'shall be of no effect' unless it falls within the powers given to the architect under the conditions and is 'confirmed in writing by the Architect within two working days of such a working direction being given'. The dangers of a contractor acting immediately on a Clerk of Works direction and something going wrong, then the architect refusing to issue an appropriate instruction, are obvious. There are, however, emergency situations where such immediate action must be taken, and the Clerk of Works' experience and skills will be required to avoid any contractual difficulties.

11.3 SUMMARY OF CLERK OF WORKS' DUTIES

In practice, the duties of the Clerk of Works are somewhat broader than those laid down in the building contract and can be summarised as including inspections, reporting in detail, advising and generally being the eyes and ears of the architect on site. The satisfactory discharge of those duties requires a wealth of practical experience, sound technical knowledge and interpersonal skills.

11.4 DUTIES OF THE CLERK OF WORKS

11.4.1 Register of drawings

A register of drawings must be maintained and all contract documentation carefully filed. In order to carry out their duties efficiently, the Clerk of Works must have an intimate knowledge of the contract drawings, the preamble, the bills of quantities and any other contract documents. The Clerk of Works should therefore ensure that there is ready access to that information, requesting this from the architect if it is not offered.

11.4.2 Errors and discrepancies

The architect should be notified of any errors in or discrepancies between the drawings, preambles, bills of quantities or other documents.

11.4.3 Decisions and variations

The architect should be notified immediately of any significant problems that arise on site, or if any decisions or variations are required. Queries should not be stored until site meetings but raised as they occur.

11.4.4 Work in accordance with the contract

The JCT Standard Building Contract states in clause 2.1 that the contractor shall carry out and complete the works in a proper and workmanlike manner, and the powers given to the architect to issue instructions in the case of non-compliance will rely on the Clerk of Works' vigilance on site.

Work should be inspected for execution in accordance with the JCT contract documents and with any instructions that may be issued from time to time by the architect. Directions to the contractor regarding defects noted should be issued in writing and a remedy sought promptly. The architect should also confirm a valid Clerk of Works direction.

The NEC supervisor ensures that the contractor compiles with the works information. The quality and quantity of the works information can determine the overall outcome of the project. The one significant contractual responsibility of the supervisor is to issue the defects certificate.

11.4.5 Materials meeting standards

Materials and goods are to be inspected for compliance with the stated standards, and to ensure that they are properly stored and protected. The Clerk of Works should inspect delivery notes, examine materials for BSI Kitemarks and CE marks, etc. and obtain necessary certificates as routine procedures.

When necessary, work in the contractor's, subcontractors' or suppliers' yards should be inspected, as should any unfixed materials and goods stored off site when requested. Any tests detailed in the contract should be witnessed and recorded.

Within the NEC any plant or materials that the works information states are to be tested or inspected before delivery to site should remain off site or not be used until the supervisor has notified the contractor that they have passed the test or inspection.

11.4.6 Clause 3.18

The Clerk of Works should confirm with the architect whether or not clause 3.18 is to operate regarding the acceptance of work or materials not in accordance with the contract. In certain circumstances the architect, with the employer's agreement and after consulting the contractor, may allow such work.

11.4.7 Directions

Directions must only be given to the main contractor through the site manager/ agent. Directions should never be given directly to workers or subcontractors' representatives. Clause 3.4 of SBC05 11 states that directions given to the contractor by the Clerk of Works shall have no effect unless confirmed in writing by the architect within two working days of such direction being given, which is extremely difficult to achieve in practice. The Clerk of Works should issue written directions, dated and numbered, with copies being passed to the main contractor, the architect, the quantity surveyor and one copy kept on site by the Clerk of Works. The NEC supervisor will issue any direction in writing, dated and numbered, with copies going to the contractor and the project manager.

11.4.8 Negligence or non-compliance

The architect should be informed immediately if there is continuing negligence or non-compliance after the contractor's attention has been drawn, both orally and in writing, to any lack of observance and there is any reason to believe that work is being sub-let without the architect's written consent as required under clauses 3.7.1 and 3.7.2.

11.4.9 Unfixed goods

Vigilance must be maintained to ensure that unfixed materials and goods are not removed from the site by the contractor, except with the architect's written approval. Refer to clause 2.24.

11.4.10 Site progress meetings

The Clerk of Works/NEC supervisor should attend site progress meetings as required and confirm the accuracy or otherwise of the contractor's progress report.

11.4.11 Daily diary

It is the duty of the Clerk of Works/NEC supervisor to keep a daily diary, recording such items as:

- temperatures, rain, snow and high winds, with particular reference to 'exceptionally adverse weather'
- records of tests carried out
- any delays that are evident, and the reasons for them
- concrete pouring and striking of formwork, etc.
- site labour and any labour problems on site
- visits by the building control officer, statutory authorities, professional team or client
- details of any condemned work or work to be covered up
- verbal instructions or information given by the architect, civil or structural engineer, or mechanical services engineer
- details of attendance at day works
- progress against programme.

Since the site diary may have to be produced when dealing with the contractor's claims, it is essential that all entries are factual and can be substantiated.

11.4.12 Weekly reports

Clerk of Works/NEC supervisor reports are to be submitted weekly in the form required by the architect or employer, and progress should be verified by recording the actual programme on the master programme.

11.4.13 Levels and setting out on new work

The Clerk of Works should verify with the contractor's representative the levels and setting out of the building. It should be noted that under clause 2.10 of SBC 11 it is the contractor's responsibility alone to set out works and establish all levels in accordance with the contract drawings, and that inspection by the Clerk of Works does not relieve the contractor of their obligation.

11.4.14 Record drawings and certification

The Clerk of Works must verify that the contractor maintains such record drawings and certifications as are required by the contract. These normally consist of internal and external drainage, electrical installation, and heating and ventilating installations.

11.4.15 Health and safety

All parties, including the Clerk of Works and NEC supervisor, should be vigilant with regard to site health and safety as per CDM 2015. Any accidents involving people, structures, equipment or plant should be recorded.

11.4.16 Materials and workmanship

Since clause 2.3 requires the contractor to carry out and complete the works using materials and workmanship of the quality and standard specified in the contract documents, it follows that the contractor alone is responsible for making certain that the works are satisfactorily completed before asking the architect to certify practical completion. It is also important, where works are specified to be to 'the reasonable satisfaction of the architect', that the Clerk of Works verifies that this standard is maintained throughout that element of the works.

The Clerk of Works should maintain systematic preliminary inspections and keep the architect informed of progress to completion and any difficulties likely to arise, and only when notified by the contractor that the works, or part of the works, are complete should the Clerk of Works then undertake a final inspection with the architect. The NEC supervisor determines if any defects are outstanding both before and after the completion certificate has been issued.

It is customary for the Clerk of Works at various stages of the works to draw the contractor's attention to items requiring corrective actions. Approaching practical completion there are often lists of defective and incomplete works being produced by the Clerk of Works, usually referred to as defects and deficiencies (snag) lists. The NEC supervisor lists any defects individually as and when they arise.

Such lists have no contractual significance under JCT contracts, and it should be noted that it is the contractor's responsibility to provide workplace supervision and ensure that proper contract information is available to enable the operatives to achieve the specified or approved standards.

Where BS EN ISO 9001:2015 is adopted by the contractor, the quality control and audit procedures may be relevant.

Clerks of Works/NEC supervisors are recommended to avoid the use of the term 'snagging lists', instead referring to 'incomplete or defective work'. Clerks of Works/ NEC supervisors should be discouraged from producing repetitive and extensive written lists of such incomplete and/or defective work, as that is the responsibility of the contractor.

The architect must make the final decision as to whether the completeness of the works is sufficient to merit a practical completion certificate within the JCT contract.

Within the NEC, the contractor corrects a notified defect before the end of the defect correction period. This defect correction period begins at practical completion for any defects notified before completion, and this tends to be between two and four weeks after completion.

If a defect is notified after completion, the employer should allow the contractor access to carry out any defects as they arise. The defects date is usually a 52-week period, and then the defects certificate is issued by the supervisor.

In some instances, the Clerk of Works may be involved in assisting the architect to prepare schedules of defects at the end of the rectification period, for rectification by the contractor prior to the issue of the certificate of making good.

11.4.17 Quality assurance

The Clerk of Works can play an important role in respect of QA procedures in contracts where BS EN ISO 9001:2015 is being applied.

11.4.18 Building control compliance

The contractor has a contractual obligation to comply with the statutory requirements, and the Clerk of Works should be aware of Building Regulations requirements and should inspect for compliance.

11.5 MATTERS BEYOND THE SCOPE OF THE CLERK OF WORKS

The Clerk of Works shall not:

- modify the design without the prior approval of the architect
- accept instructions other than from the architect
- approve the contractor's method of carrying out any remedial measures without the approval of the architect
- direct the contractor on methods of working unless agreed by the architect
- agree to commitments with subcontractors or suppliers.

11.6 SPECIFIC DUTIES

11.6.1 Day works

When the contractor contends that work, in the contractor's opinion, can only be valued on a day work basis, as opposed to measurement in accordance with the clause 13.5 valuation rules, reasonable notice of such work must be given to enable the Clerk of Works to be present and to keep records and/or sign day work sheets as the authorised representative of the architect. The sheets should be signed 'for record purposes only' since it is the architect who will decide, together with the quantity surveyor, whether or not the work is to be paid as day works. The Clerk of Works should never sign vouchers unless completely satisfied that they present a correct record of labour, plant and material; not when the Clerk of Works has not been present as a result of inadequate notification; nor when they are submitted outside the specified period in clause 5.7, i.e. not later than the end of the week following that in which the work has been executed. The NEC contract, depending on the clauses used, would require this type of work (day works) to be priced prior to the works going ahead, and this may also affect the master programme.

11.6.2 Time and material sheets

It is the practice of some contractors to submit for the Clerk of Works' signature 'time and material' sheets, for items regarded as extra or varied work. These vouchers shall not be signed by the Clerk of Works under any circumstances.

11.6.3 Liability

The Clerk of Works may be specifically appointed by the employer or the architect, or may be a permanent member of staff of the employer or the architect.

The Clerk of Works/NEC supervisor will be liable for their own acts of proven negligence, subject to the considerations of contractual liability and the vicarious liability of an employer for the actions lawfully carried out by an employee in the course of their duties influencing the outcome of any dispute involving such matters.

11.7 APPOINTMENT

The ICWCI publishes a Clerk of Works appointment document (independent practices) for use where clients wish to employ the services of self-employed Clerks of Works, and a memorandum of agreement for use in conjunction with it.

The services that may be required are listed, along with the conditions of appointment and the remuneration. The fees may be on a time charge, lump sum basis or a percentage of the total construction cost. The memorandum requires all of those facts to be recorded.

Clerks of Works are engaged under various types of contract other than JCT, and both the Clerk of Works and the NEC supervisor should acquaint themselves with the conditions of contract that will determine the extent of their powers under the contract.

Effective Reporting for a Clerk of Works/Construction Inspector Working in the Nuclear Industry

12

12.1 ASSURANCE

As the UK returns to the design and construction of nuclear projects, it is imperative that lessons are learned from history, and that quality control is tightened to ensure the highest standards are maintained.

By its nature, the nuclear construction industry can be complex, challenging and fraught with risk and uncertainty. From the early stages of building a nuclear plant, through to its construction and maintenance as part of a life cycle of facilities management, risks are high and complex. Indeed, an early pre-contract feasibility risk concerns the location of the plant. Local residents will historically resist construction and will press for strong reassurances that their homes, livelihood, environment, health and wellbeing will not be compromised.

For the Clerk of Works, this will bring added pressure, and the responsibility of ensuring that high levels of quality control are reached in a timely and consistent fashion. The Clerk of Works knows that the project will be in the public eye, and that the expectations of a nation rest on their ability as the quality control/management team to minimise risks and ensure the highest standards.

The traditional duties of a Clerk of Works/construction inspector still apply on nuclear projects – acting with ability, integrity and vigilance – but the way reporting techniques are communicated may differ from project to project. Importantly, reporting techniques need to be instantaneous, and both the contractor and the Clerk of Works should be 100 per cent conversant, confident and knowledgeable of reporting procedures prior to works commencing.

12.1.1 Ensuring high levels of quality

Traditionally, clients ensure that contract documentation contains clauses that allow the Clerk of Works to demand sample panels of brickwork and structure to be built before the work starts. For nuclear projects this is taken to the next level of quality control, and clients will insist that all areas of construction are pre-constructed as 'test bays' to ensure that any matters relating to poor quality control are resolved prior to the main works. Some clients believe this will assist contractors who may be more familiar with less stringent quality control measures to improve their skills, while other clients use this methodology to remove risk and uncertainty. No system can be 100 per cent foolproof, but this will ensure that the Clerk of Works/quality control team are confident that any issues relating to buildability and design clashes can be removed in a safe environment.

Importantly, the Clerk of Works/quality control team must develop a specialist bespoke reporting system that is fast, efficient and responsive to urgent requests for inspections of work. On large schemes, some contractors will use a zoning system and allocate specialist labour with expertise to certain sectors of the project. Others rely on software packages and BIM technology to request inspections.

For the Clerk of Works, the three principal vehicles for recording information remain site directions, the site diary and periodic reports, but the complexity of nuclear projects means that dedicated quality control management teams will receive and contextualise the information to ensure that a workable notice period (normally 48 hours) is available. The use of mobile technology and photographic evidence to support quality control is commonplace. However, because of the security requirements surrounding access to information, certain parties will control the distribution of data and information. This will not prevent the reliance and promotion of BIM technologies, but will restrict interoperability of data between all contract parties.

12.1.2 Building a quality control culture

One of the main skill sets of the Clerk of Works relates to communication skills and the ability to deliver precise, coherent messages. Importantly, where this will require a message to request removal or remediation works, the message should be consistent, professional and free from arrogance or bias.

An experienced Clerk of Works will know by instinct how and when to issue directions and notices, and will attempt to clarify the rationale and reasoning behind their decision.

In looking to develop a collaborative supportive environment, it is therefore ideal to develop a strong culture surrounding quality management and quality control systems. Changing existing cultures and practices can be extremely challenging, but most experienced Clerks of Works will identify their own methodology for success.

This does not require compromise, but a sense of shared responsibility between all parties in ensuring quality control measures are in place and observed. Questions to promote a positive culture include the following.

- Are all parties in possession of the correct drawings, specifications and details that will enable them to undertake and/or inspect the works?
- Are all parties familiar with this type of work, and do they have expertise in constructing and/or inspecting the works?
- Are all parties aware of how this section of work fits in to the main project, and the importance of its accuracy and quality?
- Are all parties aware of the consequences of not reaching high levels of quality, and the potential impact in terms of delays and their reputation?
- Are all parties willing to sign their name against this piece of work to state that they are completely satisfied that the work is appropriately complete and reaches the relevant quality control standards?
- Are all parties appreciative of the reputation that the sector has, and the importance of promoting high levels of quality control to the next generation of practitioners?

12.1.3 Non-conformity direction/instruction

Under any JCT, NEC, FIDIC or specialist form of contract, the reporting of non-conformant works should be recorded, and the nuclear industry is no exception.

Non-conformance and the use of site directions must be coherent, and each direction should record the name of the contract, project name, date of issue, sequential serial number and any relevant supporting information that will ensure the contractor and relevant parties are clear on the message. However, the added benefits of iPads or similar mobile technology will remove any uncertainty of where the non-conformity exists. The use of plain, straightforward language is important in communicating messages, and reference to supporting evidence, contract clauses, standards, illustrations, pictures or documentation will help the reader to correctly interpret messages. Copies of all communications should always be stored independently or backed up remotely.

12.1.4 Frequency of entries

Unlike on smaller projects, the frequency of reporting and the methodology of reporting techniques will be hourly and vast in number. With estimated numbers of 500-plus personnel on a typical nuclear project, it is imperative that reporting procedures are monitored on a daily basis, and that the contractor's programme for inspections, sometimes called an inspection regime, is shared between parties at a weekly or daily meeting. There should be allowance made for travel between the projects, and the parties should in all cases respect the other parties' positions.

Where modern application technology (apps) are used, it is important to ensure that diary entries have been correctly recorded. Some Clerks of Works take a secondary screenshot or picture of entries as their personal record.

12.2 The role of technology

As described earlier in this chapter, the increased use of IT systems, BIM and DIM technology and specialist software packages will revolutionise reporting/inspection techniques and will ensure that the expectations of frequency of inspections and access to updated plans, specifications and drawings are fulfilled. Indeed, during the writing of this edition, technology is being piloted that will give contract personnel access to the full range of project drawings via a visor built into a site helmet.

With the increase of mega-projects such as the ones associated with UK nuclear industries, it is certain that the use of modern digital reporting technologies such as iPads, smartphones and augmented reality technology will support and complement access to project data, ensuring that staff can access information within seconds. If unsure about the current guidelines on using technology, it is advisable to consult with a trustworthy IT specialist.

Ecological and Environmental Clerks of Works

13

An Ecological and/or Environmental Clerk of Works is employed to ensure, as far as is possible, full compliance with design standards, planning consents, permits and various statutory obligations that relate to the protection of the environment.

13.1 BACKGROUND

Across the UK, environmental regulation and planning policy impose obligations on developers and contractors to protect and enhance the environment through their work on construction sites. Very often, out of these obligations comes the need to monitor and manage the various risks arising on construction sites through the employment of specialist environmental professionals who can work alongside construction contractors.

Ecological and Environmental Clerks of Works may become involved with and be responsible for the protection and/or management of various environmental assets, including:

- ecology
- air quality
- water quality and drainage
- noise and vibration
- geology and soils (including dust control)
- landscape and visual impact
- archaeology and cultural heritage
- people and communities
- waste management
- contaminated land
- energy conservation.

13.2 PURPOSE OF AN ECOLOGICAL CLERK OF WORKS

The purpose of an Ecological and/or Environmental Clerk of Works is to:

- protect and/or manage valued ecological and other environmental features and assist in the control of risks arising from some assets; this may involve:
  - provision of practical, site-specific and proportionate assistance on how their clients can achieve compliance with relevant environmental legislation and policy encountered by the construction industry
  - advising on the onsite discharge of planning conditions and obligations imposed by local planning authorities to secure protection, management and control of environment assets consistent with local and national planning policy
  - compliance with environmental requirements set out in tender and contract documentation, environmental management systems as stipulated by clients

and the regulator, and agreements with local community groups and other stakeholders

- avoid unexpected costs, avoidable delays to project timetables, and/or adverse publicity arising from unplanned incidents, which may have future negative commercial implications and ultimately risk enforcement action and/or potential prosecution
- manage ecological and environmental operatives engaged in mitigation activities such as undertaking ecological or archaeological watching briefs, rescue or translocation of protected species, prevention of water and air pollution, and so on.

The growing demand for this role within the construction sector is demonstrated by the increasing number of ecological and engineering consultancies offering an Ecological Clerk of Works service. Also, significantly, many major infrastructure projects see the Ecological Clerk of Works as an essential role. A competent Ecological Clerk of Works can effectively oversee the management of the risks on construction sites associated with managing ecological and environmental assets, and can help to ensure a smooth and cost-efficient construction process.

Appropriately trained and qualified Ecological Clerks of Works are therefore becoming an increasingly essential part of the development process. Foreseeable increases in infrastructure development in the UK and Ireland will create a significant need for capable and competent Ecological Clerks of Works with these skills in the near future.

13.3 SCOPE

The Clerk of Works obviously has a highly valued and recognised status within the construction industry, with responsibility to ensure that work is carried out to specification, to timetable and with correct materials and standards of workmanship.

There is a corresponding need to inspect, monitor and advise on the ecological and environmental issues and constraints that arise within the construction process. This need has been emphasised by the high profile given to the Ecological Clerk of Works function in BS 42020:2013.

To ensure that this need is met in a consistently professional and competent manner, the Chartered Institute of Ecology and Environmental Management (CIEEM) has identified three roles that reflect three different situations where there is a need for the following:

- **focus on practical work:** involving practical mitigation work (such as installation of reptile or amphibian fencing) to be carried out by site 'operatives', e.g. environmental specialists or personnel employed by the site contractor
- **focus on risk management:** where an experienced ecologist or environmental

specialist takes on an advisory role, to oversee operations that may otherwise have an adverse effect on environmental assets

- **focus on quality and outcomes:** where an experienced ecologist or environmental specialist inspects works undertaken on site to ensure their compliance with design and contractual obligations, planning conditions, and/or other regulators' permits, e.g. protected species licenses.

It should be noted that the first two roles described above are likely to be appointed as an integral part of the construction contract, and are likely to be employed by the main contractor or a subcontractor. However, the third role is more likely to be employed either by the client or by appropriate environmental regulators and consenting bodies, such as local planning authorities, where these have a vested interest in compliance and the ultimate outcome of works on site.

13.4 BENEFITS ARISING FROM THE ECOLOGICAL CLERK OF WORKS ROLE

Professional Ecological Clerks of Works are able to act, for both the regulated parties and the regulators, as a mutually trusted source of advice on construction sites. They can enable developers and contractors to demonstrate commitment to compliance, earning recognition for their proactive approach to addressing environmental regulations.

Employment of an Ecological Clerk of Works provides developers and construction contractors with an effective means to demonstrate commitment to compliance with relevant environmental regulations. In this respect, consistent and regular use of an Ecological Clerk of Works presents an opportunity for 'earned recognition' with regulators, thereby reducing their need to monitor and inspect, and the likelihood of them taking enforcement action.

The Ecological Clerk of Works role offers a semi-independent source of inspection, monitoring and advice that can only improve communication and understanding between the regulators and the construction industry.

13.5 RELATIONSHIP WITH OTHER PROFESSIONALS ON SITE

To fulfil their role, it is absolutely essential that an Ecological Clerk of Works who is providing advice or practical assistance on site is an integral member of the construction team. Their role is to assist and support construction colleagues, and to provide proportionate and pragmatic solutions to environmental issues and challenges that may arise.

The ideal Ecological Clerk of Works will have an excellent understanding of all relevant ecological legislation and policy, especially as it applies to the construction environment. They will have a track record of finding innovative approaches to problem solving, and will have outstanding communication and negotiation skills.

They will also have in-depth knowledge and comprehension of what is required to achieve high ecological standards on construction sites. Working with other construction professionals and operatives, their goal will be to ensure that mitigation works are effectively implemented in a manner that meets all stated aims and objectives. Where work falls short of necessary standards and/or compliance they will be able to offer constructive recommendations for measures to address deficiencies.

To this end, the Ecological Clerk of Works will be comfortable talking with other trades and professionals, and will be able to provide site inductions and toolbox talks and ongoing advice, where required, to various work teams involved in a wide range of activities. They will be able to draw on past experience to generate innovative solutions when faced with new, complex or difficult problems.

Ultimately, their goal will be to protect and manage valued environmental features on construction sites. To this end, they will have a positive, outcome-orientated approach to their work.

Insurance Aspects

There are a number of insurances that firms should consider taking out in order to protect personal or corporate assets. Certain insurance may also be required by contract, professional body regulation or law.

The following is a brief explanation of policies commonly affected, with some observations.

14.1 PROFESSIONAL INDEMNITY INSURANCE

English common law requires the exercise of reasonable skill and care in the provision of professional services and advice that is relied on by others. The legal concept of negligence in English law is based on a failure to exercise reasonable skill and care.

Parallel and concurrent duties will also be owed by professionals in contract, even in the absence of a written contract. As such, professionals may face claims for both negligence and breach of contract from those who rely on the professional's advice and suffer a financial loss as a result. Professional indemnity (PI) insurance provides protection against such claims for negligent acts, errors or omissions in the provision of professional services. PI insurance is also referred to as 'errors and omissions' insurance, or E&O cover.

Unless a Clerk of Works is a full-time or part-time employee, it is likely they will need to arrange PI insurance themselves, regardless of whether they are working as an individual, a partnership or a limited company. Be wary of assurances such as 'Our PI insurance covers your work' without further substantiation. An employing firm may be covered under their own PI insurance for claims arising against them as a result of a sub-consultant's work, but both the firm and their insurers have a right of recovery against the sub-consultant in respect of any claims paid. The distinction between the claim against the firm, and the recovery claim from the firm against the sub-consultant, is often not recognised. As a result an unfocused question may be asked by the firm of their insurers ('Are we covered for a mistake by the sub-consultant?') and then reassurance mistakenly given to the sub-consultant ('A mistake by you is covered under our PI insurance'). Only if the insurer waives its recovery rights against the sub-consultant will the claim 'stop' with the firm, meaning that the sub-consultant is protected.

If seeking cover under another's PI, the Clerk of Works should verify that the firm's PI insurance applies in respect of sub-consultant's work (including that the professional business description is wide enough to encompass Clerk of Works services) and that a formal endorsement has been raised by the insurer, explicitly waiving its subrogation rights against them.

Due to the 'claims made' basis of PI insurance, this endorsement will need to be repeated in the PI insurance arranged by the firm for the next six to 15 years, or the protection will not exist when a claim can be made.

14.1.1 'Claims made' basis of cover

It is important to understand that PI insurance works on a 'claims made' basis of cover. This means that it is the policy to which a claim or circumstance is notified that deals with that matter, and therefore it is not the PI policy in force when the contract was signed, the work was performed, or even necessarily when the damage occurred. Notification is therefore the trigger for PI insurance. If no notification is made, or there is no policy or cover in force when the notification can be made, the claim will be uninsured. An important consideration is therefore that PI cover may have changed at the future date when the notification is made – the limit of indemnity may be lower, the excess higher or the terms of cover themselves substantially different. This is a strong argument for professionals arranging their own cover rather than relying on other firms, as they retain more control over the terms of future PI cover.

14.1.2 'Run-off' cover

A corollary to the 'claims made' basis of PI insurance is that a professional must continue to arrange PI insurance after they have ceased to practice, or they will have no cover for a future claim arising from work previously performed. This is referred to as 'run-off' insurance. Run-off insurance is typically an annual PI policy, but with a 'run-off exclusion' applied, which states that there is no cover for claims arising from work performed after a specified date (the cessation/retirement date).

Historical experience of annual run-off insurance is that in the absence of changes (whether at market level, or the level of individual risk) a professional can expect their premium to be the same in the first one to two years, and then reduce by 10 to 15 per cent year on year until the insurer's minimum premium is reached. However, as an annual contract, the insurer is free to amend the terms and premium of the insurance at each renewal.

The main consideration for the purchase of run-off insurance is statutory limitation, and therefore when the expectation would be that all claims against the professional are statute barred. This leads to the purchase of run-off for between six and 15 years.

Market practice is for run-off to remain with the insurer who insured the live firm, unless there are exceptional circumstances that dictate otherwise. However, there is only a limited market for run-off if it is not an 'own' risk.

'Block' run-off can be available. This is a run-off insurance that covers a more extended period of time (for example a six-year policy period, rather than an annual policy period). The availability of block run-off is very much subject to market factors, and is frequently unavailable at a given point in time.

14.1.3 Retroactive cover

Unless qualified by a retroactive date, a 'claims made' PI policy wording will normally provide cover for all past work performed by the insured professional.

If a retroactive date is applied, there is no cover under the policy for claims arising from work performed before the specified date. A retroactive date of 'none' is therefore the ideal position.

Typical market practice is either for a retroactive date of the inception of the firm to be applied (i.e. when it first started trading), or when the firm first purchased PI insurance.

A professional should be alert to any retroactive date applied to their insurance, and judge whether it is appropriate.

14.1.4 Breadth of cover

There is no standard policy wording in construction PI, and therefore care is needed to understand the extent of cover being provided. Professionals should ensure that they are familiar with the exclusions and conditions within their PI policy.

There are a number of standard exclusions that apply to all PI policies, including an exclusion of onerous liability accepted in contract. It is therefore important to verify that the terms of appointments, collateral warranties, novation agreements and other contracts are supported by PI insurance, and have not introduced uninsured material (very common in the construction industry).

The same principle applies to any certificates issued by a professional. An assurance that goes beyond reasonable skill may not be fully covered by PI insurance. Furthermore, any certificate issued to a third party is a form of reliance document, and the same care should be taken with these as other reliance documentation such as collateral warranties.

It should be noted that just because a document is industry standard, it is NOT automatically fully PI insurance supported. The JCT SCWa/E 2011 collateral warranty contains uninsured elements, for example.

14.2 PUBLIC LIABILITY INSURANCE

Public liability (PL) insurance provides cover for a Clerk of Works or site inspector in respect of sudden and accidental damage to other people's (third parties') property, or injury to such third parties. It is distinct from PI insurance, as PL is intended to cover damage negligently caused while providing services, as opposed to damage caused by the negligent provision of the services themselves. An example would be a Clerk of Works accidentally knocking a brick off scaffolding and it falling on a third party's car or head, causing damage or injury respectively.

For similar reasons to those detailed in the PI section above, Clerks of Works and site inspectors are advised to have their own PL insurance unless they are full-time or part-time employees and the cover is appropriately maintained by their employer. Employers should obviously have PL cover for their firm, covering all their own staff as well as any self-employed people who do not or may not maintain their own cover.

PL insurance is written on a 'claims occurring' basis, not a 'claims made' basis. This means that policy cover applies to issues arising during the policy period. Consequently future insurance cover after the services have been concluded is not the same issue as it is for PI insurance.

14.3 EMPLOYERS' LIABILITY INSURANCE

Employers' liability (EL) insurance provides cover in respect of the employer's liability for injury or illness sustained by an employee in the course of their employment. It is compulsory under the Employers' Liability (Compulsory Insurance) Act 1969, and is therefore a legal requirement for all employers.

Although the legislation is complex and detailed guidance may need to be sought, it is worth mentioning that as well as being compulsory in respect of employees, it is also compulsory for firms to have EL insurance for self-employed people who are 'employees in all but name'.

EL insurance is written on a 'claims occurring' basis.

EL insurance is distinguished from 'employers' practices' liability' (EPL), which covers employment issues (discrimination, unfair dismissal, harassment, etc.) as opposed to injury. EPL is typically provided as part of directors' and officers' insurance.

14.4 DIRECTORS' AND OFFICERS' INSURANCE

Directors' and officers' (D&O) insurance covers the principals of limited companies or limited liability partnerships. It is intended to protect individuals from claims by shareholders, regulatory authorities and other interested third parties that they have failed in their duties or responsibilities as directors or officers of the company.

An easy way to think of it is that it is like a PI policy for doing the job of being a director.

As noted above, D&O policies typically include EPL cover, to protect directors from claims by employees in relation to employment issues.

D&O insurance can also include cover for the firm, as well as the individual principals. However, policies vary and this should not be taken as a given.

14.5 CYBER LIABILITY INSURANCE

This is a recent and fast-growing area of insurance grappling with new exposures created by technology. Cover varies widely from policy to policy, but typically includes areas such as data issues (loss, theft or corruption), identification and containment of system issues (ie establishing what has been hacked and how to stop it), regulatory requirements following data breach, and defamation issues following outside interference in systems. Cover can also include loss of monies (own or third parties'), ransom attacks and loss of profit to the business due to cyber-related interruptions.

14.6 LEGAL EXPENSES INSURANCE

Legal expenses insurance provides cover for the costs involved not only in defending but, critically, in pursuing various types of legal action. As such it should be distinguished from the defence costs cover provided by PI, PL, EL, D&O or cyber policies.

Areas typically covered by a legal expenses policy include: debt recovery, contract disputes, certain investigations by HMRC, certain health and safety investigations, and statutory licence protection.

Cover varies between policies, and therefore care is needed to understand the extent of cover that is being provided. The professional should ensure that they are familiar with the exclusions and conditions within a legal expenses policy.

14.7 BUSINESS EQUIPMENT INSURANCE

This is cover for the Clerk of Works' or site inspector's own equipment, which may be lost or damaged (a laptop, for example). As this type of loss occurs outside of the home, and while being used for commercial purposes, home insurance cover may well exclude such a loss.

Specific 'all risks' cover may be arranged for such equipment. Cover varies from policy to policy, so ensure that you are familiar with the exclusions and conditions within such insurance.

It may also be possible to extend office or liability policies to include all risks cover to specified equipment.

14.8 MISCELLANEOUS

The following are common types of insurance that may be in place in connection with the project, but which would not normally be the responsibility of the Clerk of Works or site inspector.

Contractors' all risks insurance: this is a package of different covers, but one of the most important is insurance cover for the works themselves (if a fire destroys a half-built structure, the insurance is intended to cover the cost of clearing the site and rebuilding back to the same point). Under JCT design and build contracts, either the employer or the contractor is required to effect the works insurance.

Property owners' liability insurance: this is a specific form of PL policy, which deals with third-party claims against the owner of land or property in connection with their responsibilities (or default) as owners of that land or property.

Property owners' insurance: if the project relates to an existing structure, the existing owners or tenants may have arranged insurance to cover the buildings, contents, etc. of the property.

Under previous JCT design and build contracts, it was the employer's responsibility to cover the works in existing structures (Option C), and problems sometimes arose where the employer was a tenant rather than the owner, and insurance therefore already existed. The 2016 revisions to JCT design and build have sought to address this issue.

Non-negligent loss insurance: this is an optional requirement under JCT design and build clause 6.5. It is the contractor's responsibility to arrange this insurance if the option is selected. It is intended to deal with issues to adjacent property such as collapse, vibration, subsidence, changes to groundwater level and removal of support. These are areas that can be excluded under PL insurance.

Bonds and warranties: these come in many different forms, but rectification warranties are in common use in relation to construction projects. These are intended to put right problems arising, regardless of fault or legal liability. It should be noted that the underwriter of a warranty may have recovery rights against others in the event that they pay under the warranty. Hence the warranty may deal with a defect or problem, but the underwriter must then sue a professional to recover their outlay on the basis that the problem arose or was not identified sooner due to negligence of that professional. A warranty being arranged on a project reduces the PI risk to a professional (because more problems are dealt with) but does not eliminate the risk that a PI claim may be made against the professional.

14.9 PI CASE STUDIES

14.9.1 Case study 1

A self-employed Clerk of Works relies on a verbal assurance from a representative of the firm engaging them (Firm X) that their work for the firm is covered by 'their insurance', meaning both professional indemnity and public liability insurance. The Clerk of Works does not arrange their own PI or PL insurance.

A few months later the Clerk of Works fails to notice a significant departure from the architect's design and specification, which they could reasonably have been expected to notice in performing their regular inspections and supervisory duties as set out in their contract. The client subsequently incurs significant costs in rectification of the defective works, as well as delay and resulting financial losses.

While it appears, at first, that the PI insurers of Firm X are dealing with the matter, as soon as they discover that the error/omission was made by a self-employed Clerk of Works, Firm X's insurers exercise rights of recovery against the Clerk of Works (on Firm X's behalf). As the Clerk of Works has no insurance and is personally liable, the settlement is funded out of their personal assets, including savings, house, etc.

The same principles could also apply in respect of a public liability claim.

14.9.2 Case study 2

A Clerk of Works retires after a number of years trading on a freelance self-employed basis. Despite being made aware of the 'claims made' basis of cover and the need for 'run-off' cover (see section 14.1.2, above, for details) they do not take this out.

A claim is made against them a number of years after their retirement in relation to work performed prior to retirement. The fact that they had PI cover in force at the time they performed the work is irrelevant, and these policies do not respond to the claim. There is no current PI policy to which the claim can be notified, and as such it is uninsured. The Clerk of Works remains personally liable, and therefore it is only their personal assets that are available to deal with the claim – with the potentially devastating consequences outlined in case study 1.

14.9.3 Case study 3

A Clerk of Works arranges PI cover for the first time, after working on a freelance self-employed basis for almost a year without insurance. Although they are not aware of any possible claims when they arrange their PI cover, six months later they become aware of a possible claim. They notify the insurers, who point out that the policy contains a retroactive date, which is inception of the PI policy. This means that the policy only provides cover in respect of work performed after that date, so the Clerk of Works remains personally liable.

A Professional Institute

15.1 CODE OF CONDUCT

1. Election to, and acceptance of, membership of the Institute binds those so elected to this code of conduct, and also to the Institute's memorandum and articles of association. This code of conduct applies to all members, regardless of grade held, or type or status of employment, whether they continue to practise or not and whether they are in membership in the United Kingdom or elsewhere.
2. All in membership undertake to further the aims and objectives of the Institute to the best of their abilities throughout their period of membership, and to promote and advance the ethos of membership to their colleagues and fellows, and through this encourage them into membership.
3. All members shall conduct themselves in such a way that the respect, reputation, honour and dignity of the Institute is upheld, maintained and enhanced at all times.
4. All members shall carry out their professional duties to the highest standards commensurate with the Institute's motto, contained on its coat of arms, 'Potestate, probitate et vigilantia', or 'Ability, integrity and vigilance'.
5. All members shall endeavour to maintain, through CPD or other means, the highest levels of knowledge, both in spirit and deed, to ensure that they are able to comply with the requirements of law as they apply to the construction industry, including, but not restricted to, matters of health, safety and welfare as well as those that relate to the equal and fair treatment of all, regardless of creed, colour, gender, religious belief, sexual orientation or physical impediment.
6. Members shall never knowingly damage or otherwise harm the name, standing or reputation of others.

This code of conduct does not form a part of the Institute's memorandum and/or articles of association.

15.2 THE PROFESSIONAL BODY FOR THE CONSTRUCTION INSPECTORATE

Established in 1882 by a small group of Clerks of Works, the Institute of Clerks of Works and Construction Inspectorate has undergone many changes in the ensuing years. However, its name and its original principles of formation – to safeguard the requirement for ensuring quality in construction and the provision of value for money to the client – remain sacrosanct.

The Institute's members recognised that the role and duties traditionally associated with Clerks of Works are now undertaken by a far more diverse range of construction professionals, under a wide variety of job titles and, often, as an intrinsic but substantial element of another role.

In recognising this, the Institute also recognises the far wider range of construction-related disciplines that have a key role to play in the delivery of value for money to the client. The Institute firmly believes that each of these disciplines has its own specific area of expertise and knowledge that – as a progressive and learned body representing the wider Construction Inspectorate – is an essential component in the Institute's own ability to be truly representative of the discipline that it exists to promote.

15.3 BENEFITS OF MEMBERSHIP

One of the major issues that has traditionally surrounded the role of the Clerk of Works is that, on site, they are to be seen to be a benefit to the project through a fair and firm appraisal of the ongoing works. They are seen very much as 'police officers' and, in truth, that is exactly what they are. It is little wonder then that the majority will find that, although the role is probably one of the most professionally essential and satisfying in the industry, it is also isolated. The basic understanding of this particular problem was identified during the earliest stages of the formation of the Institute, and remains true even in the modern construction arena.

15.3.1 Chapters

The cornerstone of Institute membership revolves around its network of meeting centres – the Chapters – originally developed to provide a focal point for members working in isolation to meet and discuss professional issues with their contemporaries. Over the years, the network has developed to provide a presence throughout the United Kingdom, Ireland, Gibraltar and Hong Kong. Centres are strategically located to cover the following areas:

- Scotland:
  - Central
  - East
- Northern Ireland:
  - Belfast
- Ireland:
  - Dublin
- Wales
- England:
  - Devon and Cornwall
  - North East
  - Northern
  - Cumbria and North Lancashire
  - Deeside
  - North Cheshire
  - Merseyside

- o Staffordshire and District
- o East Midlands
- o Home Counties North
- o East Anglia
- o London
- o South London
- o South East
- o Southern
- o Western Counties
- o Isle of Man
- Hong Kong
- Gibraltar

The regions usually meet monthly and undertake formal CPD, visits, lectures and a wide variety of professional and social programmes as circumstances dictate.

15.3.2 Other benefits

Members enjoy a variety of other benefits, which include:

- the monthly journal/magazine *Site Recorder*, which gives details of related industry news, regulatory issues, health and safety developments, and a host of other topics relevant to the Construction Inspectorate
- regular email communication (for those linked with their local regions)
- LinkedIn group network
- access to technical advice and assistance
- access to a specialist insurance brokerage service, dealing specifically with the requirements of the Construction Inspectorate
- access to a range of annual seminars and the Institute's annual conference weekend
- biennial awards programme
- discounted ICWCI publications.

15.3.3 The biggest benefit

Membership of the Institute is recognised throughout the construction industry as the optimum level of qualification for Clerks of Works and, indeed, for those of differing titles but allied or similar functions. Not only does it demonstrate to fellow professionals that an individual has been deemed to possess the requisite expertise and knowledge to have been accepted into membership – no mean feat in itself – it also demonstrates the same unimpeachable level of professionalism to potential employers.

15.4 PRINCIPLES OF MEMBERSHIP

Membership of the Institute of Clerks of Works and Construction Inspectorate is available to suitably qualified individuals who meet the following criteria:

'They are engaged in any aspect of the inspection, superintendence of construction and/or maintenance of buildings or any other works to ensure the proper/specified use of labour and/or materials, regardless of any title given to or held by the incumbent, nor whether the role is of a practical, educational or advisory nature.'

15.5 REQUIREMENTS AND PROCEDURES

The latest requirements for gaining entry to the Institute are available either directly from the Institute's headquarters, or by email request (info@icwci.org).

15.6 ACCEPTANCE INTO MEMBERSHIP

With the exception of students (who have a separate method of entry to membership) all candidates for membership go through a structured interview procedure conducted by senior members of the Institute, occasionally assisted by relevant professionals from chartered or other bodies, depending on the discipline of the interviewee, prior to acceptance into membership.

The interview procedure will be conducted at a local meeting centre and the format of the interview will include:

- **professional practice interview:** based on the applicant's career to date, general construction-related information, health and safety, and information based on their specific discipline
- **identification of materials and their usage:** based on a selection of photographic records, candidates will be required to identify a number of construction-related materials and discuss their usage.

The interview is likely to last between one and a half and two hours.

15.6.1 Grade of membership on entry

The grade of membership offered is dependent on a recommendation being made by the interviewing panel, which itself is based on all of the information presented to the panel with each application, and the results of the interview itself.

The panel will recommend one of the three following courses of action:

- that the candidate be offered Licentiate membership (LICWCI) with a recommendation that a further interview be conducted after a period of consolidation for upgrade to full Member
- that the candidate be offered full Member grade (MICWCI)

- in the event that the members of the interviewing panel do not feel it appropriate to offer either of the above grades following interview, they may suggest Student membership and will provide a synopsis of areas that need to be developed prior to a further interview taking place, and will recommend a minimum time period before that interview occurs.

Candidates will, on completion of the interview, be given the opportunity to provide feedback to the Institute in relation to the conduct of the interview itself.

15.6.2 Election to membership

Following the deliberations of the interviewing panel, those recommended for election to Licentiate would have their names put forward to the Institute's Management Board for ratification, by no later than the end of the month following the interview.

In the case of those recommended for the immediate grade of Member, their names will be included in the next issue of the Institute's journal (*Site Recorder*) for ratification by the wider membership, again by no later than the end of the month following the interview.

15.6.3 Notification of election

Candidates are informed by letter immediately following election to membership, at which time they will be required to pay the balance of the subscriptions due for the remainder of the year, and are notified that their direct debit mandate will be activated to effect this.

Successful applicants will also be entitled to receive a Certificate of Membership at that time, which is normally presented to them at their most convenient local meeting centre. If this is not possible, alternative arrangements will be made at their entire convenience.

15.7 SUBSEQUENT UPGRADES TO MEMBERSHIP

15.7.1 Licentiate to Member

At the time that Licentiate grade is offered, the interviewing panel will also make a recommendation with regard to the minimum time that the candidate will require to develop and consolidate their knowledge prior to making an application to upgrade to Member. This will vary from one to four years, and will normally be communicated to applicants at the time that the results of their interviews are made known to them. The average period of consolidation is two years. At the time of upgrade to Member, however, all applicants must be practising as Clerks of Works or similar, and must undergo a further interview.

15.7.2 Member to Fellow

There is no interview required for upgrade to Fellow, since this level of membership recognises, subject to certain criteria, the commitment of the individual to the Institute and the discipline. Fellowship is normally available on completion of ten years in the Member grade and, at the time of application, confirmation that an individual is still practising in the role.

15.8 CONCLUSION

The Institute has a long and very well regarded history and status within the construction industry, and the designations LICWCI, MICWCI and FICWCI are accepted as the benchmark of professional recognition and standing by fellow professionals and employers alike.

For further information on the Institute, visit www.icwci.org.

Ethical and Professional Standards

16

16.1 IMPORTANCE OF ETHICAL VALUES

Ethics concern standards of behaviour that inform us how we should act in various situations, personally and professionally, that present themselves both directly and indirectly in our daily lives.

Making good ethical decisions requires a considered approach, weighing the options that impact on our course of action. Only by careful exploration of an issue or dilemma, aided by the insights and diverse perspective of others, can we make good ethical choices.

Consequently, having a systematic process for ethical decision making and the resolution of ethical dilemmas is an essential component of our professional skill set.

16.2 CWCI PROFESSIONAL AND ETHICAL STANDARDS

16.2.1 Selflessness and respect for others

Members should act solely in terms of the public interest in furtherance of their profession, and not to gain material, financial or other unfair advantage for themselves or their connections.

Members will treat others with respect, courtesy and politeness, and give due consideration to cultural differences and practices.

16.2.2 Integrity and conflict of interest

Members should not place themselves in any invidious position, either financial or otherwise, that may be perceived as compromising their integrity or their ability to act objectively, impartially or in the best interest of their client or the profession.

16.2.3 Objectivity

Members should give clear, unambiguous and appropriate advice when undertaking their professional and Institute-related duties.

16.2.4 Accountability and responsibility

Members are accountable for their actions and decisions to clients, colleagues and the public at large. Members should not transfer blame to others and will understand and act within the limits of their knowledge, ability and terms of reference.

16.2.5 Openness and transparency

Members will be truthful and transparent in all communications, sharing facts and not withholding information without reasonable cause.

16.2.6 Honesty and confidentiality

Members should act honestly, complying with relevant legislation and avoiding illegal or litigious actions. Members should at all times respect confidentiality and not divulge sensitive or restricted information unless absolutely necessary.

16.2.7 Leading by example

Members will at all times promote and demonstrate by example and good practice the profession of site inspection. They should address the interests of stakeholders in a balanced manner, ensuring that the environmental impact of their work is as positive as possible, and challenging conduct or practices that are unlawful or reprehensible.

16.2.8 Lifelong learning

Members should continue to develop professionally, maintaining and improving levels of knowledge and competence, properly recording, and demonstrating that they have undertaken at least the minimum amount of CPD as prescribed by the ICWCI.

16.2.9 Upholding the reputation of the profession and the Institute

Members should promote the Institute's vision, values and mission as the UK's leading professional body for construction site inspection, uphold the profession's good standing and refrain from conduct that detracts from its reputation. Members will also comply with all reasonable requests made by the Institute for information to support its activities.

16.2.10 Professional practice

Members should recognise and value the responsibility they have to the community, and undertake their duties with appropriate ability, integrity and vigilance.

16.3 FRAMEWORK FOR ETHICAL DECISION MAKING

The thought process below can easily be remembered by the acronym REEMA – Recognise, Establish, Evaluate, Make (a decision), Act.

16.3.1 Recognise the ethical issues

Could the situation and decision be damaging to someone or some group?
Does the decision involve a choice between a good or bad alternative, or greater or lesser evils?
Is the issue about what is legal and compliant, or what is most efficient or effective? If so, how?

16.3.2 Establish the facts

What are the relevant facts concerning the issue?
What facts are known?
Can more be learned about the situation?
Is there enough information to make a decision?
Who are the parties that have an important stake in the outcome?
Are some concerns more important than others? If so, why?
What are the options for acting?
Have all the relevant parties been consulted?
Have creative options been identified?

16.3.3 Evaluate alternative actions

Evaluate the options by asking the following questions:
Which option would do the most good and the least harm?
Which option best reflects the rights of all who have a stake?
Which option treats people most equitably and proportionally?
Which option best serves the community as a whole (not just some members)?
Which option leads you to act as the sort of person that you want to be?

16.3.4 Make a decision and test it

After considering all the approaches, which option best addresses the situation?
If you informed your family, a person you respect, or your peer group of which option you choose, what would their response be?
Reverse the perspective: ask yourself if you are comfortable with the outcome.

16.3.5 Act and reflect on the outcome

How can your decisions be implemented with the greatest care and attention to the concerns of the stakeholder?
What was the outcome of your decisions?
What have you learned from this specific situation?
How can you best consolidate and internalise the outcome?

Useful Addresses and Sources of Information

Listed below are the names and addresses of associations, institutes, government departments and other bodies from which the Clerk of Works or site inspector may seek further information.

Asbestos Removal Contractors Association (ARCA)
Unit 1 Stretton Business Park 2, Brunel Drive, Stretton, Burton upon Trent, Staffordshire DE13 0BY
Tel: 01283 566467
www.arca.org.uk

Association of Chief Estates Surveyors and Property Managers in the Public Sector (ACES)
Tel: 01257 793009
www.aces.org.uk

Association for the Conservation of Energy (ACE)
CAN Mezzanine, Old Street, 49–51 East Road, London N1 6AH
Tel: 0207 2508410
www.ukace.org

Association for Consultancy and Engineering (ACE)
Alliance House, 12 Caxton St, Westminster, London SW1H 0QL
Tel: 020 7222 6557
www.acenet.co.uk

Association of Consultant Approved Inspectors (ACAI)
Lutyens House, Billing Brook Road, Weston Favell, Northampton NN3 8NW
www.approvedinspectors.org.uk

Association of Consultant Architects (ACA)
Charlscot, Cudham Road, Tatsfield, Kent TN16 2NJ
Tel: 01959 928412
www.acarchitects.co.uk

Association for Project Safety Ltd (APS)
5 New Mart Place,
Edinburgh EH14 1RW
Tel: 0131 4426600
www.aps.org.uk

Association for Specialist Fire Protection (ASFP)
Kingsley House, Ganders Business Park, Kingsley, Bordon, Hampshire GU35 9LU
Tel: 01420 471612
www.asfp.org.uk

Brick Development Association (BDA)
The Building Centre, 26 Store Street, London WC1E 7BT
Tel: 020 7323 7030
www.brick.org.uk

British Board of Agrément (BBA)
Bucknalls Lane, Garston, Watford, Herts WD25 9BA
Tel: 01923 665300
www.bbacerts.co.uk

British Constructional Steelwork Association (BCSA)
4 Whitehall Court, Westminster, London SW1A 2ES
Tel: 020 7839 8566
www.steelconstruction.org

British Gypsum
Gotham Road, East Leake, Loughborough, Leicestershire LE12 6HX
Tel: 0115 945 1000
www.british-gypsum.com

British Precast Concrete Federation Ltd
The Old Rectory, Main Street, Glenfield, Leicestershire LE3 8DG
Tel: 0116 232 5170
www.britishprecast.org

British Standards Institution (BSI)
389 Chiswick High Road,
London W4 4AL
Tel: 0345 080 9000
www.bsigroup.com

British Woodworking Federation
The Building Centre,
26, Store Street,
London WC1E 7BT
Tel: 0844 209 2610
www.bwfcertifire.org.uk

Build UK
6–8 Bonhill Street, London EC2A 4BX
Tel: 0844 249 5351
builduk.org

Building Centre
26 Store Street, London WC1E 7BT
Tel: 0207 6924000
www.buildingcentre.co.uk

Building Engineering Services Association (BESA)
Rotherwick House, 3 Thomas More Street, St Katherine's and Wapping, London E1W 1Y2
Tel: 020 7313 4900
www.thebesa.com

Building Research Establishment (BRE)
Bucknalls Lane, Garston, Watford, Herts WD25 9XX
Tel: 0333 3218811
www.bre.co.uk

Building Safety Group (BSG)
5 Pinkers Court, Briarlands Office Park, Gloucester Road, Rudgeway, Bristol BS35 3QH
Tel: 0300 304 9070
www.bsgltd.co.uk

Building Services Research and Information Association Ltd (BSRIA)
Old Bracknell Lane West, Bracknell, Berkshire RG12 7AH
Tel: 01344 465600
www.bsria.co.uk

Carbon Trust
4th Floor, Dorset House, 27–45 Stamford Street, London SE1 9NT
Tel: 0203 9440187
www.carbontrust.com

Chartered Association of Building Engineers (CABE)
Lutyens House, Billing Brook Road, Weston Favell, Northampton NN3 8NW
Tel: 01604 404121
cbuilde.com

Chartered Association of Project Management
Ibis House, Regent Park, Summerleys Road, Princes Risborough, Bucks HP27 9LE
Tel: 0845 4581944
www.apm.org.uk

Chartered Institute of Architectural Technologists (CIAT)
397 City Road, London EC1V 1NH
Tel: 020 7278 2206
www.ciat.org.uk

Chartered Institute of Building (CIOB)
1 Arlington Square, Downshire Way, Bracknell RG12 1WA
Tel: 01344 630700
www.ciob.org

Chartered Institute of Building Service
Engineers (CIBSE)
222 Balham High Road,
London SW12 9BS
Tel: 0208 6755211
www.cibse.org

Chartered Institute of Housing (CIH)
Octavia House, Westwood Way,
Coventry, West Midlands CV4 8JP
Tel: 0247 6851700
www.cih.org

Chartered Institute of Plumbing and
Heating Engineering (CIPHE)
64 Station Lane, Hornchurch,
Essex RM12 6NB
Tel: 01708 472791
www.ciphe.org.uk

Chartered Institution of Civil
Engineering Surveyors (ICES)
Dominion House, Sibson Road, Sale,
Cheshire M33 7PP
Tel: 01619 723100
www.cices.org

Chartered Quality Institute (CQI)
2nd Floor, 10 Furnival Street,
London EC4A 1AB
Tel: 0207 2456722
www.quality.org

Concrete Society
Riverside House, 4 Meadows
Business Park, Station Approach,
Blackwater, Camberley, Surrey
GU17 9AB
Tel: 01276 607140
www.concrete.org.uk

Constructing Excellence
Bucknalls Lane, Garston,
Watford WD25 9XX
Tel: 03330 430 643
constructingexcellence.org.uk

Construction Industry Council (CIC)
The Building Centre, 26 Store Street,
London WC1E 7BT
Tel: 0207 3997400
cic.org.uk

Construction Industry Research
and Information Association
(CIRIA)
Griffin Court, 15 Long Lane,
London EC1A 9PN
Tel: 0207 5493300
www.ciria.org

Construction Industry Training Board
(CITB)
Bircham Newton, King's Lynn,
Norfolk PE31 6RH
Tel: see website for specific
departments
www.citb.co.uk

Construction Skills Certification
Scheme (CSCS)
PO Box 114, Bircham Newton,
King's Lynn, Norfolk PE31 6XD
Tel: 0344 994 4777
www.cscs.uk.com

Council on Training in Architectural
Conservation (COTAC)
The Building Crafts College, Kennard
Road, Stratford, London E15 1AH
Tel: 020 8522 1705
www.cotac.org.uk

Department for Business, Energy and
Industrial Strategy (BEIS)
1 Victoria Street, London SW1H 0ET
Tel: 020 7215 5000
www.gov.uk

Department for Environment, Food and
Rural Affairs (Defra)
Ergon House, c/o Nobel House,
17 Smith Square, London SW1P 3JR
Tel: 03459 33 55 77
www.gov.uk

Door and Hardware Federation (DHF)
42 Heath Street, Tamworth,
Staffordshire B79 7JH
Tel: 01827 52337
www.dhfonline.org.uk

Electrical Contractors' Association
(ECA)
Lincoln House, 137–143 Hammersmith
Road, London W14 0QL
Tel: 020 7313 4800
www.eca.co.uk

English Heritage
The Engine House, Fire Fly Avenue,
Swindon SN2 2EH
Tel: 0370 333 1181
www.english-heritage.org.uk

Environment Agency
Enquiries: 03708 506 506
Incident hotline: 0800 807060
Web pages accessible www.gov.uk

Federation of Master Builders (FMB)
David Croft House, 25 Ely Place,
London EC1N 6TD
Tel: 0330 333 7777
www.fmb.org.uk

Fire Protection Association (FPA)
London Road, Moreton in Marsh,
Gloucestershire GL56 0RH
Tel: 01608 812500
www.thefpa.co.uk

Fire Sector Federation (FSF)
Tel: 1608 812543
www.firesectorfederation.co.uk

Forum for the Built Environment (FBE)
Ground Floor, 4 Victoria Square,
St Albans, Hertfordshire AL1 3TF
Tel: 01625 664548
fbeonline.co.uk

Gas Safety Trust
6th Floor Dean Bradley House,
52 Horseferry Road,
London SW1P 2AF
Tel: 0207 706 5111
www.gassafetytrust.org

Gas Safe Register
PO Box 6804, Basingstoke RG24 4NB
Tel: 0800 408 5500
www.gassaferegister.co.uk

Glass and Glazing Federation (GGF)
40 Rushworth Street, London SE1 0RB
Tel: 0207 939 9101
www.ggf.org.uk

GOV.UK
www.gov.uk

Health and Safety Executive (HSE)
Various regional offices
Tel: 0300 003 1747
www.hse.gov.uk

Health and Safety Executive for
Northern Ireland (HSENI)
83 Ladas Drive, Belfast BT6 9FR
Tel: 0800 0320 121
www.hseni.gov.uk

Highways England
Various office locations
Tel: 0300 123 5000
www.highways.gov.uk

Home Builders Federation (HBF)
HBF House, 27 Broadwall,
London SE1 9PL
Tel: 0207 9601600
www.hbf.co.uk

Home and Communities Agency
Tel: 0300 1234 500
www.gov.uk

Institute of Clerks of Works and
   Construction Inspectorate (ICWCI)
Equinox, 28 Commerce Road,
Lynch Wood, Peterborough PE2 6LR
Tel: 01733 405160
www.icwci.org

Institute of Construction Management
   (ICM)
69 Adur Avenue, Shoreham by Sea,
West Sussex BN43 5NL
www.the-icm.com

Institute of Demolition Engineers (IDE)
1st, 2nd, 3rd Floor Eagle Court, 130
High Street, Rochester, Kent ME1 1JT
Tel: 01634 790548
ide.org.uk

Institute of Energy and Sustainable
   Development (IESD)
Queens Building, De Montfort
University, The Gateway,
Leicester LE1 9BH
Tel: 0116 207 8714
www.iesd.dmu.ac.uk

Institution of Civil Engineers (ICE)
1 Great George Street, Westminster,
London SW1P 3AA
Tel: 0207 2227722
www.ice.org.uk

Institution of Engineering Designers (IED)
Courtleigh, Westbury Leigh, Westbury,
Wiltshire BA13 3TA
Tel: 01373 822801
www.institution-engineering-designers.
org.uk

Institution of Engineering and
   Technology (IET)
Michael Faraday House, Six Hills Way,
Stevenage, Herts SG1 2AY
Tel: 01438 313 311
www.theiet.org

Institution of Gas Engineers and
   Managers (IGEM)
IGEM House, High Street, Kegworth,
Derby DE74 2DA
Tel: 0844 375 4436
www.igem.org.uk

Institution of Lighting Professionals (ILP)
Regent House, Regent Place, Rugby,
Warwickshire CV21 2PN
Tel: 01788 576492
www.theilp.org.uk

Institute of Revenues Rating and
   Valuation (IRRV)
Tel: 0207 831 3505
www.irrv.net

Institution of Structural Engineers
   (IStructE)
47–58 Bastwick Street,
London EC1V 3PS
Tel: 0207 2354535
www.istructe.org

Insulated Render and Cladding
   Association (INCA)
6–8 Bonhill Street, London EC2A 4BX
Tel: 0844 249 0040
www.inca-ltd.org.uk

Joint Contracts Tribunal (JCT)
28 Ely Place, London EC1N 6TD
www.jctltd.co.uk

Landscape Institute
107 Grays Inn Road, London WC1X 8TZ
Tel: 020 7685 2640
www.landscapeinstitute.org

Local Authority Builiding Control
(LABC)
3rd Floor, 66 South Lambeth Road,
London SW8 1RL
Tel: 020 7091 6860
www.labc.co.uk

Local Government Association (LGA)
18 Smith Square, Westminster,
London SW1P 3HZ
Tel: 020 7664 3000
www.local.gov.uk

Mastic Asphalt Council (MAC)
PO Box 77, Hastings TN35 4WL
Tel: 01273 242778
masticasphaltcouncil.co.uk

National Assembly for Wales
Cardiff Bay, Cardiff CF99 1NA
Tel: 0300 200 6565
www.assembly.wales

National Association of Professional
Inspectors and Testers (NAPIT)
4th Floor, Mill 3, Pleasley Vale Business
Park, Mansfield, Nottinghamshire
NG19 8RL
Tel: 0345 543 0330
www.napit.org.uk

National Building Specification (NBS)
The Old Post Office, St Nicholas Street,
Newcastle upon Tyne NE1 1RH
Tel: 0345 456 9594
www.thenbs.com

National Energy Foundation (NEF)
Davey Avenue, Knowlhill, Milton
Keynes, Buckinghamshire MK5 8NG
Tel: 01908 665555
www.nef.org.uk

National Federation of Roofing
Contractors (NFRC)
Roofing House, 31 Worship Street,
London EC2A 2DY
Tel: 020 7638 7663
www.nfrc.co.uk

National House Building Council
(NHBC)
NHBC House, Davy Avenue,
Milton Keynes, Bucks, MK5 8FP
Tel: 0800 035 6422
www.nhbc.co.uk

National Housing Federation (NHF)
Lion Court, 25 Procter Street,
London WC1V 6NY
Tel: 0207 0671010
www.housing.org.uk

National Inspection Council for
Electrical Installation Contracting
(NICEIC)
Warwick House, Houghton Hall Park,
Houghton Regis, Dunstable LU5 5ZX
Tel: 0333 202 5720
www.niceic.com

National Insulation Association (NIA)
2 Vimy Court, Vimy Road,
Leighton Buzzard LU7 1FG
Tel: 08451 636363
www.nia-uk.org

National Trust
Heelis, Kemble Drive,
Swindon SN2 2NA
Tel: 0344 800 1895
www.nationaltrust.org.uk

NEC Contracts
One Great George Street,
London, SW1P 3AA
Tel: 0207 6652446
www.neccontract.com

Neighbourhood Energy Action (NEA)
West One, Forth Banks,
Newcastle upon Tyne NE1 3PA
Tel: 0191 261 5677
www.nea.org.uk

Painting and Decorating Association (PDA)
32 Coton Road, Nuneaton,
Warwickshire CV11 5TW
Tel: 0247 6353776
www.paintingdecoratingassociation.co.uk

Property Care Association
11 Ramsay Court, Kingfisher Way,
Hinchingbrooke Business Park,
Huntingdon PE29 6FY
Tel: 0844 375 4301
www.property-care.org

Public Sector Audit Appointments (PSAA)
PSAA Limited, 3rd floor, Local
Government House, 18 Smith Square,
London SW1P 3HZ
Tel: 020 7072 7445
www.psaa.co.uk

Royal Incorporation of Architects in Scotland (RIAS)
15 Rutland Square, Edinburgh EH1 2BE
Tel: 0131 2297545
www.rias.org.uk

Royal Institute of British Architects (RIBA)
66 Portland Place, London W1B 1AD
Tel: 0207 5805533
www.architecture.com

Royal Institute of Chartered Surveyors (RICS)
12 Great George Street,
London SW1P 3AD
Tel: 024 7686 8555

www.rics.org

Royal Town Planning Institute (RTPI)
41 Botolph Lane, London EC3R 8DL
Tel: 0207 9299494
www.rtpi.org.uk

Scottish Government
St Andrew's House, Regent Road,
Edinburgh EH1 3DG
Tel: 0300 244 4000
www.gov.scot

Society for the Protection of Ancient Buildings (SPAB)
37 Spital Square, London E1 6DY
Tel: 0207 3771644
www.spab.org.uk

Timber Trade Federation (TTF)
The Building Centre, 26 Store Street,
London WC1E 7BT
Tel: 020 3205 0067
www.ttf.co.uk

Town and Country Planning Association (TCPA)
17 Carlton House Terrace,
London SW1Y 5AS
Tel: 0207 9308903
www.tcpa.org.uk

Timber Research and Development Association (TRADA)
Stocking Lane, Hughenden Valley, High
Wycombe, Buckinghamshire HP14 4ND
Tel: 01494 569600
www.trada.co.uk

Water Regulations Advisory Scheme (WRAS)
Unit 13, Willow Road, Pen y Fan
Industrial Estate, Crumlin,
Gwent NP11 4EG
Tel: 0333 207 9030
www.wras.co.uk

Chapter 17

Wood Panel Industries Federation (WPIF)
Autumn Park Business Centre, Dysart
Road, Grantham, Lincolnshire NG31 7EU
Tel: 01476 512 381
www.wpif.org.uk

Index

Index

Index

Index